Urban Agriculture and City Sustainability

Urban Agriculture and City Sustainability

Editors

S. Syngellakis
Wessex Institute, UK

J. L Miralles i Garcia
Politechnic University of Valencia, Spain

WIT PRESS Southampton, Boston

Editors:

S. Syngellakis
Wessex Institute, UK

J. L Miralles i Garcia
Politechnic University of Valencia, Spain

Published by

WIT Press
Ashurst Lodge, Ashurst, Southampton, SO40 7AA, UK
Tel: 44 (0) 238 029 3223; Fax: 44 (0) 238 029 2853
E-Mail: witpress@witpress.com
http://www.witpress.com

For USA, Canada and Mexico

Computational Mechanics International Inc
25 Bridge Street, Billerica, MA 01821, USA
Tel: 978 667 5841; Fax: 978 667 7582
E-Mail: infousa@witpress.com
http://www.witpress.com

British Library Cataloguing-in-Publication Data

A Catalogue record for this book is available
from the British Library

Library of Congress Catalog Card Number: 2019932596

ISBN: 978-1-78466-365-0
eISBN: 978-1-78466-366-7

The texts of the papers in this volume were set individually by the authors or under their supervision. Only minor corrections to the text may have been carried out by the publisher.

The material contained herein is reprinted from issues of the International Journal of Design & Nature and Ecodynamics, Volume 13, Number 4, Volume 14, Numbers 1 and 2, published by WIT Press.

Preface

This volume contains papers presented at the 1st International Conference on Urban Agriculture and City Sustainability in Southampton, UK, organised by the Wessex Institute, UK and the Universitat Politècnica de València, Spain. The conference was sponsored by the WIT Transactions on Ecology and the Environment, the International Journal of Sustainable Development and Planning and the International Journal of Design and Nature and Ecodynamics.

This book is a result of the first meeting in Southampton, UK in 2018. The challenge of the sustainable city is more important for our society every day and every year. Every year the population in the cities grows in the world. Food production and food sovereignty are essentials for sustainability in the cities. We are living at a time when researchers from all over the world are looking for new forms of food production in cities. Urban agriculture becomes progressively another way of understanding the city. Thereby the city is transformed by innovative rewilding processes.

Papers from the conference, as well as those contained in this volume, have been archived in the Wessex Institute digital library (www.witpress.com/elibrary), where they are permanently available to the international community. The success of the first conference rests on the need to continually improve urban agriculture systems, as well as rewilding cities. Urban agriculture includes a lot of multidisciplinary approaches to a new perspective of the city like: urban metabolism, green infrastructure, productive urban landscape, peri-urban agriculture, city food chains, market gardens from development and planning point of view; organic farming, patchwork farms, roof top farms, green roofs, vertical greening and farms, hydro-aqua-aeroponics systems, green houses or food networks from urban farms point of view; community cohesion and job creation, community gardens, co-producers or education and training from socio-economics point of view; and transport and distribution, waste food recovery, healthy cities, care farming, biofuel production, street quality, eco-cities or air pollution reduction from innovation and benefits. The development of appropriate urban agricultures systems is one of the essential concepts to achieve sustainable cities and a way to improve the food sovereignty.

Urban agriculture is a research field that represents a great challenge but also a great opportunity to improve the green infrastructure and the production of healthy food in our cities and their surroundings.

The Editors are grateful to the members of the International Scientific Advisory Committee and other colleagues who helped to review the papers, as well as all authors for their significant contributions.

The Editors
2019

Contents

URBAN FOODPRINTS (UF) – ESTABLISHING BASELINE SCENARIOS FOR THE SUSTAINABILITY ASSESSMENT OF HIGH-YIELD URBAN AGRICULTURE

K. BENIS[1], R. GASHGARI[2], A. ALSAATI[2] & C. REINHART[1]
[1] Massachusetts Institute of Technology, Department of Architecture, Cambridge USA.
[2] Center for Complex Engineering Systems, KACST, Riyadh, Saudi Arabia.

ABSTRACT

Allowing for significant water savings and year-round yields, Controlled-Environment Agriculture (CEA) is oftentimes portrayed as a sustainable alternative to conventional farming, and its practice in urban areas as a food, income and employment generator is expanding worldwide. Particularly in today's fast growing cities, where economic strength is buying food security through imports, a large-scale implementation of such practices should be further investigated as potential contributors – not only to food security but also to self-sufficiency – for the production of horticultural crops. However, further than quantifying the potential for food self-sufficiency of cities through urban cultivation, there is a crucial need for assessing the extent to which such scenarios are effectively more sustainable than existing supply chains. For that purpose, this paper presents the *Urban Foodprints (UF)* methodology, a fundamental preliminary step in the sustainability assessment of high-yield urban agriculture, consisting of collecting and integrating data on the existing supply chain, to be used as a baseline scenario in the environmental performance analysis. Through the case of Riyadh, Saudi Arabia, where harsh climatic conditions, a heavy reliance on food imports and a growing population constitute major threats to food security, the UF method is described and applied to the top four consumed horticultural crops – watermelon, tomato, onion and carrot. The environmental sustainability of high-yield urban agriculture in Riyadh is subsequently assessed for tomato, as a comparison of the resulting city's current *foodprint* for the crop vs. a scenario of local production in CEA urban farms. Results show that urban production in high-yield greenhouses has the potential to reduce Global Warming Potential (GWP) by 9%. However, while water savings contribute greatly to reducing irrigation-related emissions and food miles are considerably reduced, the energy needs of the greenhouses are significantly higher than the baseline. This outcome may be improved by enhancing the envelope of the farms to reduce overheating.

Keywords: baseline scenario, Controlled-Environment Agriculture (CEA), sustainability assessment, Urban Foodprint, urban food system.

1 INTRODUCTION

While our cities are growing at an unprecedented pace, their relationship with the environment has deeply changed, as low fossil fuel prices and sprawling transportation infrastructures have been connecting them to increasingly remote and global hinterlands. The latter have been fueling the growing resource use intensity of urban areas, which have in turn been releasing more and more waste and emissions into oceans and the atmosphere. As a result, today's cities are heavily reliant on large-scale provision of food that is traveling from further and further away before reaching the urbanites' plates. With emerging sustainability concerns, the environmental footprint of these complex food systems, including cultivation, processing and transport, has been increasingly assessed by the scientific community over the past decade [1], [2]. At the same time, as part of their sustainability agendas, cities are promoting urban food production as a means to reduce the demand for agricultural land elsewhere and shorten food miles [3]. To rate the extend of urban food production, two metrics are being used, self-sufficiency (or self-reliance) [4], [5] and food security. According to the United Nations *'food self-sufficiency is generally taken to mean the extent to which a country can*

DOI: 10.2495/DNE-V13-N4-349-360

satisfy its food needs from its own domestic production' [6]. Food security on the other hand occurs *'when all people, at all times, have physical and economic access to sufficient, safe and nutritious food that meets their dietary needs and food preferences for an active and healthy life'* [7]. Food self-sufficiency is therefore a measure of national food independence whereas food security is concerned with the stability and affordability of food supply chains. The latter is thus more applicable for cities that tend to seek local self-reliance to gain economic and social benefits, including job creation, increased property values, and community empowerment [8].

Urban self-sufficiency has been investigated by several researchers. In Singapore, looking exclusively at public housing properties, researchers have assessed the country's potential for vegetables self-sufficiency through the large-scale implementation of high-yield roof-integrated farms. Their results show that a nationwide deployment of such systems could satisfy 35.5% of domestic demand [9]. In the United States, another study estimated the level of food self-sufficiency of Cleveland, as an example for a typical post-industrial North American city. The food categories that were considered were fresh fruits and vegetables, eggs, poultry, and honey. The three assessed scenarios led to overall levels of self-reliance between 4.2% and 17.7% by weight and 1.8% and 7.3% by expenditure in total food and beverage consumption, compared to the current level of 0.1% self-reliance in total food and beverage by expenditure. The authors concluded that significant levels of local self-reliance in food, the most basic need, is possible in this type of cities, which have been plagued with home foreclosures and resulting vacant land, lack of access to healthy food, hunger, and obesity especially in disadvantaged neighborhoods [4]. Similarly, various scenarios were assessed to measure the potential level of self-sufficiency for vegetables in Montréal, showing that the city could easily satisfy its entire vegetable demand, both through high-yield hydroponic roof-integrated systems, and through low-tech on-soil farming [5]. In Europe, researchers quantified the potential of rooftop vegetable production in the city of Bologna, showing that a city-wide implementation of rooftop greenhouses could satisfy 77% of local demand [10]. In Lisbon, it was estimated that, if all the available vacant land within the Metropolitan Area was used for high-yield soilless farming, 124% of the demand for vegetables could be met locally [11]. All these case studies are showing the potential of cities for self-reliance for fresh produce, under current crop yields and consumers' demand, simply by using their vacant areas.

From an environmental sustainability standpoint, some assessments have shown that increasing self-sufficiency through local production for some crops is not necessarily more sustainable than current practice. In the UK, a study showed that producing greenhouse strawberries in London may have a higher carbon footprint than importing Spanish greenhouse strawberries [12]; in Austria, researchers found that imported tomatoes from Spain and Italy have two times lower Greenhouse Gas (GHG) emissions than those produced nationally in capital-intensive heated systems [13]. Therefore, further than quantifying the potential for food self-sufficiency of cities through urban cultivation, there is a crucial need for assessing the extent to which such scenarios are more resource efficient than existing supply chains. For that purpose, a fundamental step in the sustainability assessment of alternative local food supply practices is the assessment of existing supply chains, to be used as baseline scenarios of the analysis. In this study, we will refer to these baseline scenarios as *'Urban Foodprints'* (UF), a concept that has been previously used to express resource consumption and environmental impacts associated with the urban food system, from agricultural production to distribution and consumption (e.g. [14]).

This paper is organized as follows: Section 2 synthesizes the existing literature on environmental sustainability assessment of urban CEA and identifies research gaps. Section 3 describes a new methodology to establish baseline scenarios for the sustainability assessment of urban food production followed by an application of the method to Riyadh, Saudi Arabia as a case study in section 4.

2 ASSESSING ENVIRONMENTAL IMPACTS OF URBAN CEA

The role of urban agriculture – as a source of local and fresh food – in enhancing urban food security in cities has been widely recognized [15] and its commercial-scale implementation in highly populated urban areas has led to an increasing interest over the past decade [16]. By *commercial urban agriculture*, here we refer to high-yield food production in Controlled-Environment Agriculture (CEA) units such as Vertical Farms (VF), Rooftop Greenhouses (RG) or Shipping Container Farms (SC) as defined in a previous study [16], located within the urban built environment. Oftentimes portrayed as sustainable alternatives to conventional food production, these commercial farms are gaining momentum, both among entrepreneurs and academics. Allowing for significant water savings and year-round yields and fostered by technology developments such as highly efficient spectrum-specific grow lights and computer-assisted climate and crop control systems, the practice of CEA as a food, income and employment generator is expanding in major cities worldwide. However, while there is a large body of literature quantitatively evaluating crop growth in controlled environments such as greenhouses, the sustainability assessment of CEA facilities within urban contexts is a relatively new field of research [16].

Since Despommier's aspirational depiction of the vertical farm concept was published almost a decade ago [17], CEA farms have sprouted in several cities around the globe and several attempts have been made to quantify how CEA's overall environmental impact compares to current agricultural practices. In London, a study estimated the environmental footprint of lettuce production in a hypothetical high-rise urban vertical farm and compared it to the environmental footprint of lettuce conventionally grown in the UK in winter and summer [18]. Similarly, other authors have compared resource use intensity of CEA in cities like Barcelona, New York and Boston, to conventional cultivation facilities in respective countries, for the production of tomato and lettuce [19], [20]. While these approaches compare the environmental performance of the growing processes involved in energy-intensive CEA vs. conventional cultivation, they do not offer a complete assessment of whether indoor farming could mitigate environmental impacts of the vegetables that are currently distributed in cities. To do so, it would have been necessary to compare the footprint of the urban farms to the footprint of the existing supply chain for the vegetables, i.e. to include all the – domestic and foreign – locations of origin of the tomato and lettuce, measuring environmental impacts of respective production modes in each one of these locations, and finally considering conditioned transportation of the produce from the farms to the city.

In a recently published article, two of the authors of this study performed such an assessment, comparing high-yield production of tomatoes in conditioned urban farms in Lisbon vs. the existing supply chain for tomatoes distributed in the city [11]. Establishing the baseline scenario involved extensive data collection on all the locations from which tomatoes are being sourced, including farming practices, average yields, resource use, and food miles. Similarly, baseline scenarios were built for tomato supply in three additional cities of different sizes, located under different climate conditions and with diverse foodshed characteristics – Singapore, Paris and New York – and the environmental footprint of their current supply chains for tomatoes were compared to hypothetical urban rooftop-integrated conditioned greenhouses and closed shipping container farms. Whereas rooftop greenhouses were found to significantly reduce GHG emissions under all four climates, shipping container farms only yielded a positive environmental balance vis-à-vis the baseline in megacities located in colder climates, that seasonally rely on long distance food imports [21]. Whereas Despommier presented urban vertical farms as the absolute sustainable solution to the world's food supply problems, these outcomes reveal how context-specific the sustainability of urban food production is, and highlight the importance of building baseline scenarios that take into account the existing supply chain for the assessed crops in each particular urban context.

Establishing these scenarios is a data-intensive task and collecting the data can be challenging. Next section describes the UF methodology that we used to build these baselines, including major challenges and pitfalls.

3 THE URBAN FOODPRINTS METHOD

Defining *Urban Foodprints* consists of getting snapshots of the existing food system for a given urban area, using metrics related to food demand, resource use intensity of production, and food miles, to estimate the overall environmental impacts caused by the supply of a given produce to the city. The following sections describe the steps of the data collection to build the *Urban Foodprint* of a given crop, for a given urban area.

3.1 System boundary and origin of produce

The system boundary defines the phases of the process that will and will not be included in the assessment. Life cycle assessments of food products are usually conducted from cradle to farm gate, focusing on the most environmentally-harmful phase of the products' lifespan, cultivation [1]. To enable comparisons of the existing supply chain with scenarios of local food production, *Urban Foodprints* encompass a larger set of activities, from cradle to distribution, in such a way as to include travel distances and transportation modes from farm gate to the assessed urban area (see Fig. 1).

Data to determine the share of the supplied crop that comes from domestic production and the share that comes from imports can be found in official agriculture and trade statistics reports and websites from national departments of agriculture and trade. Within domestic production, defining the regions of the country where the crop is produced allows to calculate

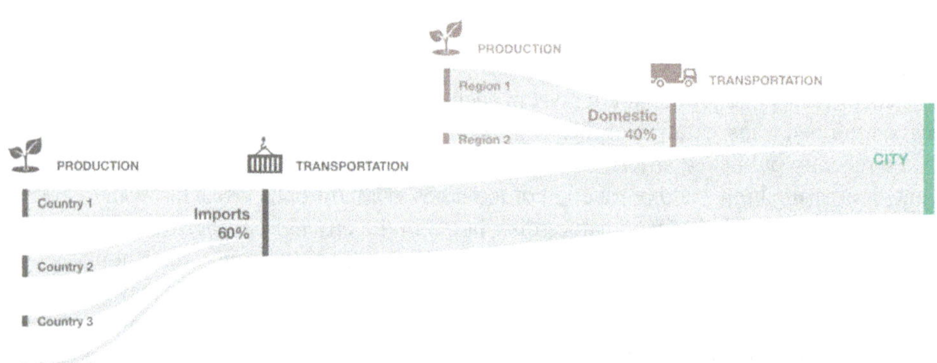

Figure 1: System boundary of the *Urban Foodprints* method.

average travel distances of the crops from respective regions of origin to the city. Similarly, imported produce is disaggregated among its countries of origin, based on official trade and customs data. Food miles can subsequently be calculated as follows:

$$FM_{AV} = \sum_{i=1}^{n} w_i FM_i \#(1)$$

where FM_{AV} (average *Food Miles*) is the weighted average distance travelled by the crop from its location of production to the city (tkm); w_i is the relative weight of each origin of supply; and FM_i are the respective food miles travelled from each location of supply (tkm), which vary according to the food system of the city that is being assessed.

3.2 Yields

Furthermore, gauging to what extent high-yield urban farms can be more efficient than their conventional counterparts requires comparing respective yields, i.e. amounts of produce grown per unit of area per year in each case. For that purpose, based on the collected data on origin of produce (section 3.1), it is necessary to further investigate the dominant cultivation conditions for each location including local climate conditions, whether the crop is produced in open fields or in protected environments as well as the length of the growing season. In case the crop is produced indoors, knowing whether it is grown on-soil or through soilless cultivation techniques such as hydroponics will further inform us on the potential yields. Such information can be found in national reports, websites of national departments of agriculture, and scientific articles. Unfortunately, it can be challenging to find this information for all the regions and countries from which the crop is sourced, especially when assessing a crop whose countries of origin have a restricted or underdeveloped data infrastructure. When no data is found, assumptions have to be made, based on crops with similar growth requirements, or data from neighboring countries with similar climatic conditions and farming practices.

Once this additional data collection phase is completed, based on respective shares of origins of the assessed crop (as defined in section 3.1) and on the farming techniques that are practiced in these different locations, average yields of existing supply chains are calculated as follows:

$$YIELD_{AV} = \sum_{i=1}^{n} w_i YIELD_i \#(2)$$

where $YIELD_{AV}$ is the weighted average yield of the crop currently supplied to the city (kg/m^2); w_i is the relative weight of each origin of supply; and $YIELD_i$ are the respective crop yields for each location (kg/m^2), which vary according to local climatic and technological conditions in the farms.

3.3 Resource use

In addition to the type of farming methods practiced in different countries, UF also requires data on the amount of water use for irrigation as well as average energy use for climate control (in case of conditioned greenhouse farming) and machinery. This information may be available in national reports, websites of national departments of agriculture, and scientific articles. Finding this data can be challenging especially for (1) crop types that have attracted less attention within the scientific community or (2) countries for which data is generally scarce or difficult to access. Again, when no data is found, assumptions have to be made, based on crops with similar growth requirements, or data from neighbouring countries with similar climatic conditions and farming practices.

3.4 Global warming potential

Finally, greenhouse gas (GHG) emissions related to irrigation, operational energy and transportation of the existing supply chain of the assessed crop to the city is calculated (expressed in $kgCO_2eq/kg$), using the relevant emission factors, as follows:

$$GWP = \sum_{i=1}^{n} w_i \left(WU_i \times E_i^{IR} + EU_i \times E_i^{EL} + F_i \times E_i^{TR} \right) \#(3)$$

where *GWP* is the Global Warming Potential of the crop currently supplied to the city ($kgCO_2eq/kg$); w_i is the relative weight of each origin of supply (defined in section 3.1); WU_i is the water use per kilogram of produce (l/kg); EU_i is the energy use per kilogram of produce (kWh/kg); F_i represents the freight transport of one ton of produce over a distance of one kilometre (tkm); and E_i^{IR}, E_i^{EL} and E_i^{TR} are the respective emission factors of irrigation, electricity generation and refrigerated transportation for each origin of supply.

4 THE RIYADH CASE-STUDY

This section presents the application of the above laid out UF method to Saudi Arabia's capital city, Riyadh, and discusses the results.

4.1 Background

Today, less than 2% of land is arable in Saudi Arabia, a desert country with no permanent lakes, depleted aquifers and very little precipitation, where overgrazing and intensive agriculture are further accelerating land degradation and desertification. Surprisingly – in spite of its harsh climate and severe groundwater depletion – the kingdom produces and exports poultry, eggs, dairy products, fish, dates, fruits and vegetables. Food and water security is presently considered high as the country can rely on its oil-based economy to import most of the food and to desalinate water [22]. However, maintaining this security will be a key challenge in the coming decades due to Saudi Arabia's growing population and steady urbanization rate: the kingdom's population, 32.3 million in 2016, is expected to grow by 85% by 2050 [23] and the percentage of urbanites, 83% as of 2014, is growing at an annual rate of 2% [22]. By 2050, the country is expected to import practically all food from abroad [24].

Following the galloping urbanization trend, Riyadh's population has been growing steadily over the past decades – making it the most populous city in Saudi Arabia with more than 8 million residents in 2017 – and is expected to further increase by 53% by 2075 [25]. The city is located in the major agricultural region of the country, where 49% of the kingdom's vegetables are grown and occupy over half of the total national horticultural production area [26]. Major crops include tomatoes, cucumber, eggplant, onion, watermelon, squash and pumpkin, and depend heavily on irrigation, accounting for 88% of national water use in 2009 [27]. In this context of inefficiency amidst resource scarcity, the government is now promoting greenhouse cultivation and drip irrigation as more sustainable farming practices. As of

2013, Riyadh region accounts for more than 50% of greenhouse production of horticultural crops [26].

As mentioned above, CEA uses significantly less water than conventional farming, does not rely on arable land and is independent of ambient climatic conditions [16]. It therefore seems to be a particularly attractive technology for Saudi Arabia, where oil-based (and therefore vulnerable) economic strength is currently buying food and water security. Large-scale CEA implementation could therefore contribute not only to the country's food security but also to self-sufficiency with horticultural crops, which represent one third of the Saudi diet [28]. This case study explores to what extend CEA could contribute to Riyadh's self-sufficiency for the city's four major horticultural crops – tomatoes, carrots, watermelons and onions.

4.2 Urban foodprints of main horticultural crops

Figure 2 presents the UF of the four main horticultural crops supplied to the city of Riyadh, and the following subsections comment on the data collection process. Figure 3 shows the example of data collection for the urban foodprint of tomato supplied to Riyadh.

4.2.1 Origin of produce

Data on origin of produce were mainly collected from the statistics of the Food and Agriculture Organization of the United Nations [28] for 2013. Watermelon and tomato are among the

		Watermelon	Tomato		Onion		Carrot	
1. Origin of produce *(a)*	Domestic [%]	100	79		25		67	
	Imports [%]	0	21		75		33	
			Jordan	11	Egypt	40	Australia	40
			Syria	3	Yemen	29	China	29
			Egypt	3	India	5	Turkey	5
			Turkey	3				
	Avg. food miles *(b)* [km]	1290	1362		1642		3267	
2. Yields	[kg/m2]	2.2 *(c)*	5.9 *(f)*		1.9 *(a)*		1.8 *(c)*	
3. Resources	Water [l/kg]	449 *(d)*	104 *(g)*		323 *(h)*		394 *(g)*	
	Energy [kWh/kg]	0.33 *(e)*	0.91 *(g)*		2.50 *(i)*		1.13 *(g)*	
4.GWP *(j)*	[kgCO2eq/kg]	22.74	6.31		13.05		19.71	

(a) According to [28];
(b) Calculated, based on average distances between ports and then to Riyadh;
(c) Calculated from [26];
(d) According to [29];
(e) Calculated from [30];
(f) Calculated from [24];
(g) Calculated from [31];
(h) According to [32];
(i) According to [33];
(j) Calculated from [34], [35] and [36].

Figure 2: Urban foodprints of the four main horticultural crops supplied to Riyadh.

major horticultural crops produced in Saudi Arabia, and thus most of the produce supplied to Riyadh comes from domestic production, whereas imports are more significant for onions and carrots.

To calculate food miles, global trade routes were used as a reference. The ports with the highest activity intensities were selected. Land and sea were the main transport modes for most of the countries of origin except for Australia, where air is the main transportation mode to Saudi Arabia. Jeddah Islamic Seaport is the biggest and busiest port in the Middle East and North Africa (MENA) region. Consequently, it is considered as the main destination port for countries that ship containers to Saudi Arabia. As for land ports, Durra Border Crossing and Halat Ammar are the main ports that connect Eastern Mediterranean countries to Saudi Arabia. Approximate distances from the exporting countries to the ports were calculated, and subsequently the distances of moving the crops from the ports to Riyadh were measured. Average food miles were then estimated.

4.2.2 Yields

The existing literature was surveyed to determine shares of the supply produced in open fields vs. shares produced in greenhouses (see Fig. 3 for the case of tomatoes), and respective yields under each condition in respective locations of origin. While it is extremely important to find recent data that depicts as accurately as possible the current situation, for some of the cases, the most recent available data were found in studies published around a decade ago. In the meantime, diverse contextual factors such as technological improvements in greenhouse energy management or efficiency improvements in irrigation systems may have had relatively important impacts on yields.

4.2.3 Resource use

Estimating energy and water use is the most challenging step of establishing the foodprint of a crop, as such data are not available in FAO datasets, and has therefore to be collected from scientific articles. When no published articles for the specific crop or for its production in a

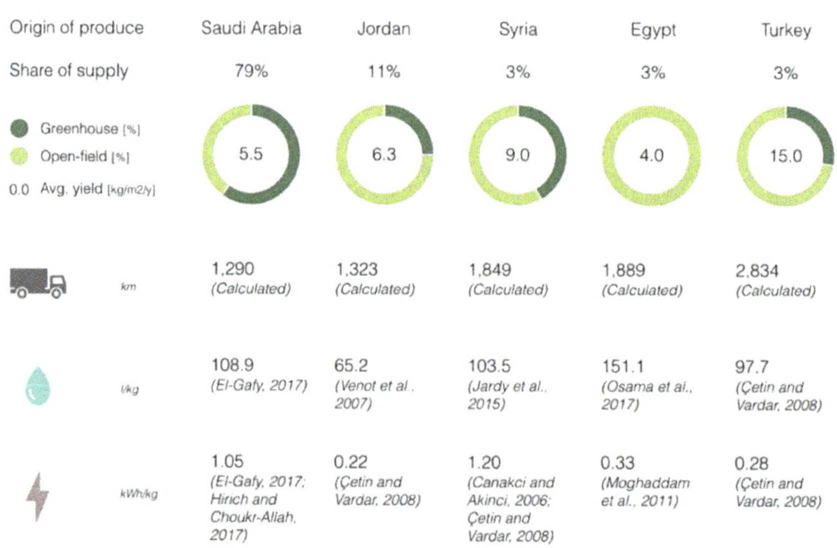

Figure 3: Urban foodprint of tomato supplied to Riyadh.

specific country were found, data were collected for similar crops, i.e. from the same family, having similar growth requirements, or from countries with similar climatic and technological conditions.

4.2.4 Global warming potential
Finally, GHG emissions related to water, energy and transportation were calculated based on relevant energy mixes and emission factors.

4.3 Comparison to local production – the case of tomato

In an alternative scenario of urban food production in Riyadh, we focused on tomato, the crop with the lowest calculated baseline GWP (see Fig. 2), to gauge whether local production could perform better. All the fresh tomato was assumed to be produced in rooftop greenhouses within the metropolitan area, traveling an average distance of only 30 km to reach distribution points. A simulation workflow for crop production in urban environments – previously presented and applied to four cities [11], [21] – was applied here. Riyadh weather data were input to the model, and construction properties of the hypothetical greenhouses were defined according to the most widespread properties of existing facilities, i.e. a steel structure with a polycarbonate cover, NFT hydroponic equipment, backup lighting and an HVAC system set to maintain indoor temperatures within the optimal range for tomato growth [21]. Figure 4 synthesizes results.

Whereas commercial farming in hydroponic greenhouses can lead to efficiency gains of a factor of 8 in terms of yields and a factor of 2.7 in terms of water use, operational energy use per kilogram of produce is over four times higher in such facilities than in the baseline scenario. This is mainly due to the high cooling needs of the greenhouses under Riyadh's hot desert climate. When converting water use, operational energy and transportation to GHG emissions, results showed that urban production in high-yield greenhouses has the potential to reduce GWP by 9%. While water savings contribute greatly to reducing irrigation-related emissions, the energy needs of conditioned hydroponic greenhouses are significantly higher than the baseline (where 50% of the tomatoes come from open field cultivation, where energy requirements are drastically lower). Finally, transportation is considerably reduced. In spite

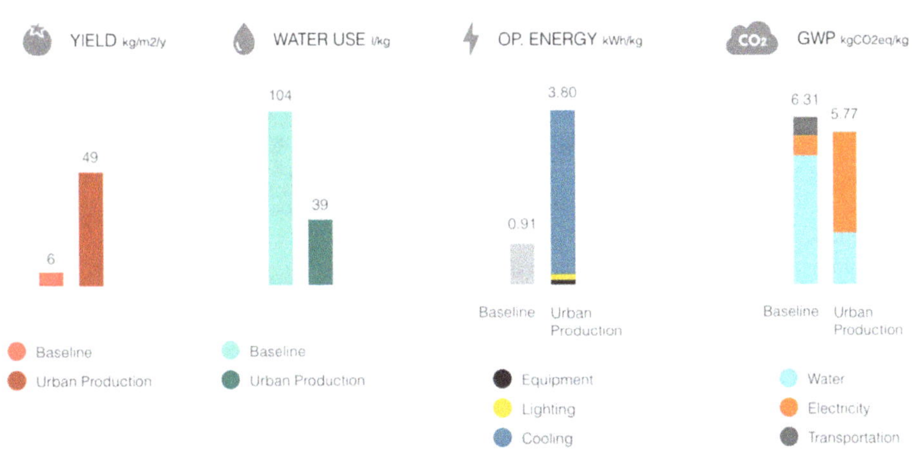

Figure 4: Baseline vs. urban production scenario.

of the positive impact on GWP, this result is significantly below potential environmental impacts mitigations estimated for other cities, such as Lisbon – where a potential GWP reduction of 50% was found [21] – located in a more temperate climate, where greenhouse conditioning is less energy-intensive.

5 CONCLUSIONS

This article described the *Urban Foodprints* method for the elaboration of baseline scenarios – an essential preliminary step in sustainability assessment of urban CEA scenarios. The method was applied to the city of Riyadh, Saudi Arabia. The main obstacles and challenges that were encountered throughout the data collection process were mentioned, being mainly related to the scarcity of reference sources, or, when data were not available, to accuracy in the process of making assumptions. In fact, while establishing a baseline scenario is crucial for any sustainability assessment, up-to-date and consistent data on food systems can be extremely challenging to find, as it was the case for some of the crops supplied to Riyadh. Yet even when no data are available, when thoroughly completed with grounded assumptions, UF can provide a close approximation to the baseline situation and therefore constitute a reliable starting point for sustainability assessments of urban food production scenarios.

ACKNOWLEDGEMENTS

Generous support for this work has been provided by the King Abdulaziz City for Science and Technology (KACST) as part of the research under the Center for Complex Engineering Systems (CCES) at MIT and KACST through the Urban Water, Energy and Food project.

REFERENCES

[1] Benis, K. & Ferrão, P., Potential mitigation of the environmental impacts of food systems through Urban and Periurban Agriculture (UPA) – A life cycle assessment approach. *Journal of Cleaner Production*, **140(2)**, pp. 784–795, 2017. https://doi.org/10.1016/j.jclepro.2016.05.176

[2] Notarnicola, B., Tassielli, G., Renzulli, P.A., 3Castellani, V. & Sala, S., Environmental impacts of food consumption in Europe. *Journal of Cleaner Production*, **140(2)**, pp. 753–765, 2017.

[3] Baker, L. & de Zeeuw, H., Urban food policies and programmes. In *Cities and Agriculture: Developing Resilient Urban Food Systems*, eds. H. de Zeeuw & P. Dreschel, Earthscan from Routledge, pp. 26–55, 2015.

[4] Grewal, S.S. & Grewal, P.S., Can cities become self-reliant in food? *Cities*, **29(1)**, pp. 1–11, 2012. https://doi.org/10.1016/j.cities.2011.06.003

[5] Haberman, D., Gillies, L., Canter, A., Rinner, V., Pancrazi, L. & Martellozzo, F., The potential of urban agriculture in Montreal: a quantitative assessment. *ISPRS International Journal of Geo-Information*, **3(3)**, pp. 1101–1117, 2014. https://doi.org/10.3390/ijgi3031101

[6] Thomson, A. & Metz, M., *Implications of Economic Policy for Food Security: A Training Manual*, Food and Agriculture Organization of the United Nations: Rome, 1998.

[7] Rome Declaration on World Food Security and World Summit Plan of Action, Food and Agriculture Organization of the United Nations, November 1996, Rome, Italy. http://www.fao.org/docrep/003/w3613e/w3613e00.HTM (accessed on 11 June, 2018).

[8] Mok, H.F., Williamson, V.G., Grove, J.R., Burry, K., Barker, S.F. & Hamilton, A.J., Strawberry fields forever? Urban agriculture in developed countries: A review.

Agronomy for Sustainable Development, **34(1)**, pp. 21–43, 2014. https://doi.org/10.1007/s13593-013-0156-7

[9] Astee, L.Y. & Kishnani, N.T., Building-Integrated Agriculture: Utilising rooftops for sustainable food crop cultivation in Singapore. *Journal of Green Building*, **5(2)**, pp. 105–113, 2010. https://doi.org/10.3992/jgb.5.2.105

[10] Orsini, F., Gasperi, D., Marchetti, L., Piovene, C., Draghetti, S., Ramazzotti, S., Bazzocchi, G. & Gianquito, G., Exploring the production capacity of rooftop gardens (RTGs) in urban agriculture: the potential impact on food and nutrition security, biodiversity and other ecosystem services in the city of Bologna. *Food Security*, **6(6)**, pp. 781–792, 2014. https://doi.org/10.1007/s12571-014-0389-6

[11] Benis, K., Reinhart, C. & Ferrão, P., Development of a simulation-based decision support workflow for the implementation of Building-Integrated Agriculture (BIA) in urban contexts. *Journal of Cleaner Production*, **147**, pp. 589–602, 2017. https://doi.org/10.1016/j.jclepro.2017.01.130

[12] Kulak, M., Graves, A. & Chatterton, J., Reducing greenhouse gas emissions with urban agriculture: A life cycle assessment perspective. *Landscape and urban planning*, **111**, pp. 68–78, 2013. https://doi.org/10.1016/j.landurbplan.2012.11.007

[13] Theurl, M.C., Haberl, H., Erb, K.-H. & Lindenthal, T., Contrasted greenhouse gas emissions from local versus long-range tomato production. *Agronomy for Sustainable Development*, **34(3)**, pp. 593–602, 2013. https://doi.org/10.1007/s13593-013-0171-8

[14] Goldstein, B., Birkved, M., Fernandez, J. & Hauschild, M., Surveying the environmental footprint of urban food consumption. *Journal of Industrial Ecology*, **21(1)**, pp. 151–165, 2016. https://doi.org/10.1111/jiec.12384

[15] Eigenbrod, C. & Gruda, N., Urban vegetable for food security in cities. A review. *Agronomy for Sustainable Development*, **35(2)**, pp. 483–498, 2014. https://doi.org/10.1007/s13593-014-0273-y

[16] Benis, K. & Ferrão, P., Commercial farming within the urban built environment – Taking stock of an evolving field in northern countries. *Global Food Security*, **17**, pp. 30–37, 2018. https://doi.org/10.1016/j.gfs.2018.03.005

[17] Despommier, D., *The Vertical Farm: Feeding the world in the 21st century*, Thomas Durne Books: New York, 2010.

[18] Al-Chalabi, M., Vertical farming: skyscraper sustainability? *Sustainable Cities and Society*, **18**, pp. 74–77, 2015. https://doi.org/10.1016/j.scs.2015.06.003

[19] Sanyé–Mengual, E., Oliver-Solà, J., Montero, J.I. & Rierdevall, J., An environmental and economic life cycle assessment of rooftop greenhouse (RTG) implementation in Barcelona, Spain. Assessing new forms of urban agriculture from the greenhouse structure to the final product level. *International Journal of Life Cycle Assessment*, **20(3)**, pp. 350–366, 2015. https://doi.org/10.1007/s11367-014-0836-9

[20] Goldstein, B., Hauschild, M., Fernandez, J. & Birkved, M., Testing the environmental performance of urban agriculture as a food supply in northern climates. *Journal of Cleaner Production*, **135**, pp. 984–994, 2016. https://doi.org/10.1016/j.jclepro.2016.07.004

[21] Benis, K, Reinhart, C. & Ferrão, P., Building-Integrated Agriculture (BIA) in urban contexts: Testing a simulation-based decision support workflow. *Proceedings of the 15th IBPSA Conference*, pp. 1942–1951, 2017.

[22] Food and water security in the Kingdom of Saudi Arabia; Lovelle, M., Strategic Analysis Paper, Future Directions International, 28 July, 2015. http://www.futuredirections.

org.au/publication/food-and-water-security-in-the-kingdom-of-saudi-arabia/ (accessed on 14 June, 2018).

[23] Saudi Arabia in World Population Ageing 1950–2050; United Nations, Population Division, DESA, pp. 402–403, 2016. http://www.un.org/esa/population/publications/worldageing19502050/pdf/177saudi.pdf (accessed on 14 June, 2018).

[24] Fiaz, S., Noor, M.A. & Aldosri, F.O., Achieving food security in the Kingdom of Saudi Arabia through innovation: Potential role of agricultural extension. *Journal of the Saudi Society of Agricultural Sciences*, article in press, 2016. https://doi.org/10.1016/j.jssas.2016.09.001

[25] Statistical Year Book for 2017; General Authority for Statistics. https://www.stats.gov.sa/en/930 (accessed on 4 July, 2018).

[26] Statistical Year Book for 2013; General Authority for Statistics. https://www.stats.gov.sa/en/46 (accessed on 15 June, 2018)..

[27] Irrigation in the Middle East region in figures, AQUASTAT Survey 2008, FAO, 2009. http://www.fao.org/docrep/012/i0936e/i0936e00.htm (accessed on 14 June, 2018).

[28] Agricultural statistics database of the Food and Agriculture Organization of the United Nations, FAOSTAT, 2013. http://www.fao.org/faostat/en/#data (accessed on 15 June, 2018).

[29] Policy Options for Reducing Water for Agriculture in Saudi Arabia, KAPSARC, 2016. https://www.kapsarc.org/wp-content/uploads/2016/04/KS-1630-DP024A-Policy-Options-for-Reducing-Water-for-Agriculture-in-SA.pdf (accessed on 15 June, 2018).

[30] Moradi, R., Moghaddam, P.R. & Mansoori, H., Energy use and economical analysis of seedy watermelon production for different irrigation systems in Iran. *Energy Reports*, **1**, pp. 36–42, 2015. https://doi.org/10.1016/j.egyr.2014.10.002

[31] El-Gafy, I., Water-food-energy nexus index: analysis of water-energy-food nexus of crop's production system applying the indicators approach. *Applied Water Science*, **7(6)**, pp. 2857–2868, 2017. https://doi.org/10.1007/s13201-017-0551-3

[32] Mahmoud, M.A. & El-Baby, A.Z., Crop water requirements and irrigation efficiencies in Egypt. In *The Handbook of Environmental Chemistry*, Springer, Berlin, Heidelberg, 2017.

[33] Allali, K., Dhehibi, B., Kassam, S. & Aw-Hassan, A.A., Energy consumption in onion and potato production within the Province of El Hajeb (Morocco): towards energy use efficiency in commercialized vegetable production. *Journal of Agricultural Science*, **9(1)**, pp. 118–127, 2017. https://doi.org/10.5539/jas.v9n1p118

[34] Daccache, A., Ciurana, J.S., Rodriguez Diaz, J.S. & Knox, J.W., Water and energy footprint of irrigated agriculture in the Mediterranean region. *Environmental Research Letters*, **9(12)**, p. 124014, 2014. https://doi.org/10.1088/1748-9326/9/12/124014

[35] Brander, M., Sood, A., Wylie, C., Haughton, A. & Lovell, J., Technical Paper: Electricity-specific emission factors for grid electricity, Econometrica, 2011. https://ecometrica.com/assets/Electricity-specific-emission-factors-for-grid-electricity.pdf (accessed on 15 June, 2018).

[36] Wang, J., Rothausen, S.G.S.A., Conway, D., Zhang, L., Xiong, W., Holman, I.P. & Li, Y., China's water-energy nexus: greenhouse-gas emissions from groundwater use for agriculture. *Environmental Research Letters*, **7(1)**, p. 014035, 2012. https://doi.org/10.1088/1748-9326/7/1/014035

NEW POLICIES FOR THE MANAGEMENT OF PERI-URBAN AGRICULTURAL SPACES. THE CASE OF L'HORTA DE VALÈNCIA (SPAIN)

J.L. MIRALLES I GARCIA
Department of Urban Planning, Polytechnic University of Valencia, Spain.

ABSTRACT

On March 12, 2018, the Law of L'Horta de València was approved. This law is the result of a very long process of technical and social maturity, of more than 20 years. The peri-urban agricultural area of L'Horta de València has historical, environmental, economic and social values that have been recognized for a long time and currently occupies some 23,000 ha. However, the protection established in territorial planning is a passive protection only for land uses control, but does not include actions or initiatives to maintain agricultural activity which is precisely the object of protection. In the year 2000, a popular legislative initiative collected 118,000 signatures in favour of the approval of a law to protect L'Horta. This law was not approved, but it opened a period of consolidation of public awareness for the protection of this space and analysis of the active measures to be taken to ensure the maintenance of agricultural activity in a manner compatible with the landscape, as well as environmental and historical qualities. All this process, in which a large number of social agents have intervened, has finally culminated in the Law approved this year. The present article synthesizes the contributions made on the subject and analyses the actions that the law finally regulates as instruments for the protection and revitalization of this agricultural space. Particularly the article analyses its general approach; the elements to protect that includes especially the agricultural economic activity; the legal framework of the territorial plan; the treatment of abandoned lands through measures such as the 'land bank'; zoning and classification of agricultural land including measures for the treatment of urban edges; the creation of a Council of L'Horta de València to manage this area; the definition of the Agrarian Development Plan; and the implementation of the General Register of Agricultural Production.
Keywords: agricultural heritage, agricultural land management, peri-urban zones management, regional planning, Valencia's huerta.

1 INTRODUCTION

La Huerta is an extraordinary space, as evidenced by many studies about all its characteristics. All these studies have been carried out over a long period. For example, the classic that Cavanilles [1] did in the late eighteenth century to others studies made in the 90s like those of Miralles [2], Díaz [3], Antolin [4] and Biot [5]. All the studies have in common the positive evaluation of this singular landscape.

Antolín already points out that the lands of high agrological capacity in the Valencian Country represents 3.9% of the surface of the Valencian territory. The more agricultural land are the fertile plain of Turia river in Valencia and the fertile plain of the lower valley of Segura river on the south of Alacant. At the same time, in 2006, the urbanized areas suppose a 4.86% of the Valencian territory.

Really, these agricultural zones are rare. At the international level, the Dobris Report [6] highlighted the importance and uniqueness of this landscape by identifying only six similar locations throughout Europe.

Currently, the most complete and rigorous study is the document named Territorial Action Plan to regulate and revitalize L'Horta de Valencia (*Pla d'Acció Territorial d'Ordenació i Dinamització de L'Horta de València,* in Catalan), promoted and elaborated since 2006 by the regional Ministry of Environment, Water, Urbanism and Housing of the Autonomous Government of the

DOI: 10.2495/DNE-V13-N4-361-372

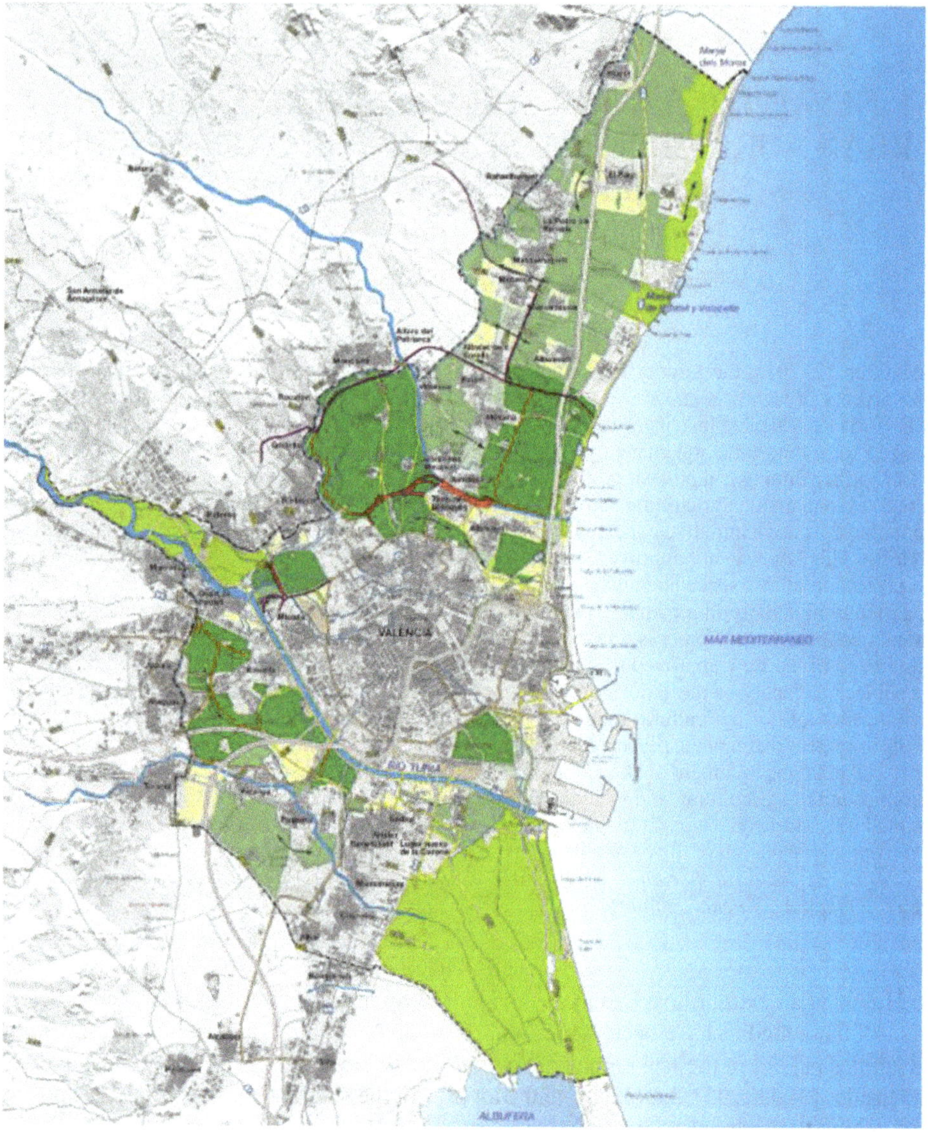

Figure 1: Geographical area of the peri-urban agricultural space of L'Horta de Valencia and zoning. Origin: Territorial Action Plan to regulate and revitalize L'Horta de Valencia. Generalitat Valenciana.

Generalitat Valenciana (*Conselleria de Medi Ambient, Aigua, Urbanisme i Habitatge*, today named *Conselleria d'Habitatge, Obres Públiques i Vertebració del Territori*). The plan was exposed to the public for first time in 2010 and after stopped. You can see in Fig. 1 the agricultural area of L'Horta de Valencia according to this plan. In 2015 was revised and again exposed to the public twice more in 2016 and 2017. Now, the plan is in final revision to approval.

On the other hand, the urban expansion on Mediterranean coast of Spain started by the Economic Plan of 1959. In 60s, public administration promoted a big touristic resort in the area named El Saler of the Valencian coast, similar to the project of La Manga del Mar Menor, in Murcia (Spain) [8][9][10][11]. This project generated a very strong social movement against the destruction of spaces with environmental values like La Albufera, El Saler, the agricultural spaces of L'Horta and the old bed of the Turia River [12].

This social movement grew while the destruction of agricultural land also grew by urban development [13]. In 2001, this social movement promoted a law to protect the agricultural land of L'Horta by a popular legislative initiative procedure. The text of this proposal was the result of different workshops to analyze the problems and propose actions. The initiative managed to collect 118,000 signatures in favor of the law but it was rejected by regional Valencian parliament.

However, since this time, many studies started about the situation of agricultural land of L'Horta, its problems and possible actions to maintain this anthropic landscape. Particularly the studies to elaborate a territorial plan of L'Horta landscape, already mentioned. Finally, on 6 March, 2018, the 'L'Horta law' (*Llei de L'Horta de València,* in Catalan, or *Ley de la Huerta de Valencia,* in Spanish) was passed and published on 12 March, 2018.

2 L'HORTA'S LAND: SYNTHETIC RELATION OF PROBLEMS AND MANAGEMENT ACTIONS

2.1 Main problems identified on agricultural land of L'Horta de València

Many authors studied different questions about the peri-urban agricultural space of L'Horta de Valencia. This section includes, without intending to be exhaustive, the main authors who have worked on the subject of L'Horta from different points of view.

Muñoz [14], [15] carried out a complete study promoted by Valencian Government as a part of first works to elaborate the territorial plan. More recently, Romero & Francés [16] coordinated a work with an update of global diagnosis.

In addition, other authors studied different particular questions about this space:

- Guinot & Esquilache [17] and Temes & Moya [18], [19] studied the evolution and dynamics of agricultural landscape changes.
- The collective book 'Farmhouses: Landscape and architecture of L'Horta' [20] study the architectural and public works heritages. Note you that the oldest farmhouses that today exist are Gothic. In addition, the irrigation network is from the Muslim time or older.
- Miralles, Díaz & Altur [21], Miralles [22] and Gaja [23] studied the speculative urban development on Mediterranean coast of Valencia.
- Miralles [24], [25] and Romero & Melo [26] studied the processes of planning and management of peri-urban agricultural land of L'Horta.
- García & Cabrejas [27] and Marqués, Segura & Maroto [28] studied sociological questions like the perception from people about the social and environmental benefits made by peri-urban agricultural landscapes.

Finally, in 2017, Valencia organized the *2017 MILAN URBAN FOOD POLICY PACT Annual Gathering and Mayor's Summit.* One of the results was the document 'Proposals for a transition to food governance' [29].

The studies carried out are of very diverse type and by a large number of authors over a long period. It can be said that both the diagnosis and the management proposal regulated in the law, are the result of an immense collective work discussed on innumerable occasions. All this activity has produced documents of all kinds and has progressively generated a collective awareness that little by little has fixed into a set of concrete proposals.

2.2 Main actions identified to solutions

The first studies with the specific issue of protection and management the agricultural land of L'Horta was carried out to elaborate the text of law to protect L'Horta by popular legislative initiative in 2001 [30]. Afterwards, the initial analysis and proposals improved with the contributions of different authors and new studies like already cited.

Next, the main analyzes and proposals because of this historical process are described in a synthetic way.

2.2.1 Moratorium of new urban development and regional plan to agricultural land of L'Horta

The studies showed that an accelerated process of urban development progressively urbanized an important part of the peri-urban agricultural territory (urban uses, facilities and infrastructures). In fact, approximately 50% of the historic agricultural land has already disappeared due to urban expansion processes. It is therefore necessary to carry out a spatial planning plan, called 'territorial action plan' in Valencian legal terminology, to order urban uses, establish protected areas and green infrastructure and approve a moratorium on urbanization processes of agricultural soils while drafting the plan.

2.2.2 Reconversion of agriculture to improve profitability: agricultural plan and agricultural council

The landscape of L'Horta is an exceptional landscape with a special feature: it is an anthropic landscape. It is a landscape generated by the accumulated wisdom for centuries by farmers. This implies an essential conclusion: the maintenance of this space is only possible if we have an agricultural activity compatible with its values and, additionally, maintaining the agricultural activity so that the farmers can live with dignity of their work.

This vision implies foreseeing, allowing and regulating actions to improve the agricultural production and the management of the production with a global vision and perspective more integrated in the processes of the global economy. It is possible promote a large list of actions: direct sales by farmers; proximity markets; certificate of origin; farmer reconversion to use new technologies such as the sale of organic products over the Internet; management center for the agricultural spaces and their production, coordinated investment in production and infrastructures...

2.2.3 Metropolitan plan and new governance

In addition to the territorial action plan of L'Horta, it is also necessary to draw up a plan for the metropolitan area of Valencia. This plan is necessary to design the location of urban uses, infrastructures and strategic facilities at the metropolitan level, assuming the green infrastructure proposal that includes the agricultural spaces of L'Horta. Through a metropolitan plan, urban expansion can be reoriented towards other areas. The elaboration of this plan is complex and also its management or execution.

Nevertheless, if we want planning the metropolitan area it is necessary to address the issue of metropolitan governance. In Spain, it exits the local and the regional administration. The metropolitan level is intermediate between the municipal and regional levels. Obviously, for the plan to be successful, it is necessary to foresee the management system for the execution the actions planned. In the Valencian case, the Autonomous Government has the competences for the elaboration and approval of the metropolitan planning, but the urbanistic competences are municipal. Therefore, the metropolitan plan establishes the overall vision; the municipalities develop the metropolitan plan in its territory each administrative department executes, where appropriate, the infrastructures and facilities planned. In any case, inter-administrative coordination is not easy. For example, certain infrastructures are a state competence and, in these cases, the regional government cannot impose its design to the central administration.

2.2.4 Environmental power to manage natural Capital

From the analyzes carried out, it arise the idea of generating an institution of environmental power with the objective of guaranteeing in the long term the natural resources that are considered necessary for future generations and that, therefore, constitute an unrepeatable Natural Capital. This institution or organization can be called the Natural Capital Bank. This institution must be independent of the other powers of states: legislative, executive and judicial. The idea is the following: the legislative power establishes the protections and the environmental power is responsible of the custody of territory to maintain the naturals resources protected according to protected rules. It is a way to block protected natural resources.

Now, in the Spanish and Valencian case, the implementation of this action implies the modification of the Spanish Constitution or the Statute of Valencian Autonomy. It is not possible to implement this measure from the current administrative organization and legal regulation or only partially and without full effectiveness.

2.2.5 Sustainability costs should be society's costs: agricultural council

On the one hand, in order to launch all kinds of strategic and investment agricultural actions with a global vision, it is necessary to create a board or council to manage revenues and expenses with those objectives. This council must have the responsibility of manage the agricultural plan.

On the other hand, the need for sustainability produces a cost for today, to secure long-term benefits. These costs of securing long-term benefits should be distributed among the beneficiaries of environmental services. If general society is the beneficiary, then it should bear the cost. A management system could be created based on land stewardship contracts inside the agricultural plan financed by private or ecological tax. The council must have also the responsibility to manage these possibilities.

2.2.6 Revised property theory about the ground to build on

The current legal system of Spain and the Valencian Country, facilitates the processes of real estate speculation and land speculation generating real estate bubbles that produce dramatic economic crises, as well as the irreversible destruction of spaces with environmental value such as the peri-urban agricultural spaces.

On the other hand, it is evident that the speculative processes based on changing the sale value of land and buildings do not generate economic progress but, on the contrary, poverty. In Spain, three real estate bubbles with the corresponding crises have actually occurred.

To protect agricultural land it is necessary changing this situation. In practice, the change must go in the line of distinguishing between ground and underground and consider that the underground, which support the foundation of building and civil works, is a natural asset and therefore a public good.

Now, the regulation of these aspects corresponds to the central administration of Spain and the Autonomous Communities only can complete central regulation. Therefore, this issue cannot be addressed directly with regional legislation alone.

3 MANAGEMENT SYSTEM FOR AGRICULTURAL LAND IN NEW LAW OF L'HORTA

As has been mentioned, the Law of L'Horta is the result of a long process of collective maturation of the Valencian society on the problems that affect the emblematic agricultural productive area of L'Horta. The Law is not the result of the work of a particular technical team. Rather, the technical team that written the final text worked to give coherence, to compile and to formalize contributions made from many authors and forums. The same law has been submitted to a process of public participation. In this process, the stakeholders done contributions to the text. We must remember that environmental regulations are subject to the process of strategic environmental assessment and, therefore, before its legislative procedure, it goes through a process of public participation about the law's text.

It can be said that the idea of approving a Law of L'Horta is old, but the precision of its contents began to materialize with the text of the law to protection L'Horta promoted (but not approved) by popular legislative initiative in 2001. The Law now approved and published in the Official Gazette of the Generalitat Valenciana on March 12, 2018 [31] is the inheriting of that popular legislative attempt.

The purpose of this law is *the preservation, recovery and revitalization of the Huerta as a space with recognized agrarian, environmental, landscape, architectural, historical, cultural and anthropological values, which are decisive for economic progress, the quality of life of the citizens and the sustainable management of the metropolitan area of Valencia, promoting the economic profitability and viability of the agricultural activity.*

The initial articles of law establish the general framework and, among them, highlight the recognition of the social and public function of the agricultural land. Thus, it considers that the agrarian activity and the natural, cultural and landscape patrimony of the Huerta de Valencia have a relevant social role by favoring the development of the agrarian sector, food sovereignty, the welfare of the people, the sustainable use of the territory and the prevention of climate change. This means in practice a new perception of peri-urban agricultural activity. This declaration of its social function also means, within the Spanish legal framework, the declaration of public utility and social interest of all the actions that are carried out, which allows, where appropriate, the expropriation.

This new perception also includes the possibility of land stewardship contracts or similar to manage the agricultural activity or, in general, agreements between owners, farmers and consumers or users for the planning and management whichever of the values of the natural, cultural and landscape patrimony of L'Horta.

The main innovations that the law contribute are comment in the following sections.

3.1 Elements that make up the Huerta

The law describe the elements in a broad sense, not just physical elements. Therefore, the law establishes the following list:

- People who dedicate themselves to agriculture professionally.
- The Water Court of Valencia and its historic irrigation communities, the Royal Canal of Montcada and the rest of the irrigation communities.

- The land of high agrological capacity.
- The hydraulic heritage and water.
- The architectural, archaeological and ethnological heritage.
- The natural heritage (ecosystem).
- The agricultural network roads and historical roads.
- The structure and plot system of the Huerta de Valencia.
- Agricultural activity.
- Any element material or immaterial nature whose maintenance is adequate in order to promote the distinguishing feature and the sense of belonging.

From this list, we highlight as a first element the professional farmer, that is, the farmer who has the agricultural activity as a means of life. We must bear in mind that this peri-urban agricultural area is smallholder, that is, there is a large number of people with little plots for agricultural exploitation of less than one ha. Although it is common that farmers have several plots of land owned for exploitation. Obviously, to maintain this agricultural space, the maintenance of farmers and their activity is essential.

Another noteworthy aspect is the Water Court [32], inscribed in 2009 on the Representative List of the Intangible Cultural Heritage by the UNESCO. The Court resolves, only by verbal argument, conflicts between irrigators since medieval times. Farmers accept and apply the rulings of the Court. This system, in current terminology, is in fact a mediation system.

3.2 Territorial action plan for planning and promoting the Huerta of Valencia

It is evident that it is necessary a spatial regional plan for regulate the uses in the peri-urban agricultural area of L'Horta de València. The plan must identify the different areas and environmental zones, as well as the landscape units. The law includes the elaboration of territorial plan already started in 2006 and today almost ready for final approval (see section1).

The plan has the following objectives: establish protected areas with their justification and specific regulations for uses in each area; establish areas where urban expansion is possible; and the provision and limitations of public facilities and infrastructures, especially those of transport. All this is important and, in our case, the agricultural space of L'Horta becomes a green metropolitan infrastructure because it constitutes a natural resource and a landscape to protect and establishes limitations to other uses that may involve the destruction or disappearance of more agricultural land.

However, the territorial plans establish a framework of uses, but they do not have the capacity to regulate or promote economic, social or environmental activities, nor for the financial or economic management of activities. Therefore, the territorial plan must be complemented by other types of measures.

The law and, in more detail, the territorial plan establish the permitted uses, mainly agricultural, compatible uses as public facilities and tertiary compatibles with agriculture activity and prohibited or very limited uses such as industrial, urban expansion or highway infrastructure.

3.3 Agricultural land neglected

The law proposes intervention instruments in abandoned or uncultivated agricultural land. The legislator aims to maintain agricultural activity in the protected land and, for this, establishes the possibility of temporary transfer of use of the plot through an agricultural leasing, which allows its cultivation by third parties who do not own the land. In this case, professional farmers have preference to be beneficiaries of the leases.

3.4 Classification and zoning of the Huerta's land

The Valencian urban legal system classify lands into urban (currently built), urbanizable (building land) and non-urbanizable (land designated as not for building) and, in turn, the non-urbanizable land can be protected or not protected.

On the other hand, zoning allows the delimitation of specific zones within each land class with specific characteristics and its different regulatory treatment.

In this general framework, the law introduces the special urban treatment of the areas named 'Huerta recovery place' and the 'Huerta recovery sector'. These areas are usually borders zones between urban land and protected land. The law establishes a special management system for these areas.

The 'recovery places' are little geographic zones, usually on urban edges, with dilapidated buildings and sealed soil. In these zones, the law establishes a special treatment: building rights are recognized for all the plots but accumulating the building in a part of the area and recovering at least 2/3 of the surface for agricultural cultivation. It is a system to design the urban areas on its edges with agricultural land.

The 'recovery sectors' are geographic areas, usually also on urban edges, but larger than the previous ones where there are dilapidated buildings and degraded agricultural land. In these areas, it is expected to accumulate the building in 1/3 of the surface and allocate the other 2/3 to crops in plots that are assigned to the owners who are beneficiaries of the 'reparcelling' project (project to change initial plots of an area to new building plots with other design). The law foresees an agricultural leases system in case the owners do not cultivate the land.

3.5 L'Horta council

It is one of the key pieces for the management of the peri-urban agricultural space. This council is conceived as an entity manager assigned to the Regional Ministry with competencies in agriculture. The council has the legal form of consortium with autonomy to take initiatives and actions and manage its own assets for the fulfillment of its purposes.

The council has a wide range of functions, among which highlight the followings:

- Elaborate the agrarian development plan.
- Acquire or lease land (for example, to cultivate).
- Obtain economic resources and manage them (for example, from the European Union).
- Mediation to facilitate transfer of agricultural uses land or lease agricultural parcels. Manage a land bank to agricultural uses.
- Manage the quality brands of products in the area.
- Manage a rural guards service.
- Maintenance of roads and ditches.
- Agricultural professional training.
- Agricultural research and development.
- Prepare mandatory and binding reports on actions in the field of L'Horta.
- Establish collaboration agreements with other entities.
- Perceive and manage fees.

The L'Horta Council is an essential entity for the promotion of agricultural activity and the maintenance of the historical landscape. It must act as a catalyst and promote all those actions of all kinds that help a more competitive agricultural activity while maintaining its own landscape within the framework of a globalized economy.

3.6 The agrarian development plan

It is another of the key elements because of agricultural production cannot reach its optimum based solely on the individual action of farmers. A global vision is necessary and to design and execute key strategic actions. The L'Horta Council has the competence to elaborate the plan that today is drawing up.

The plan has among its main objectives the followings:

- The improvement of agricultural infrastructure, such as irrigation networks and roads, and rural security.
- The professionalization and improvement of the viability and profitability of agricultural holdings.
- The improvement of quality productions, including quality brands, consumption, proximity, agro-ecology, sustainability, the use of local varieties.
- The improvement and implementation of marketing and distribution networks, the promotion of short marketing channels or direct sales from farmers.
- The diversification of agricultural incomes through the rational and limited implementation of complementary uses and activities compatible with the main agricultural activity.
- The mechanisms of management and intermediation of huerta lands, destined, in a special way, towards people who dedicate themselves to agriculture professionally, with preference to young people and women.
- Generational shift.
- The development of hiring formulas such as the land stewardship contracts, payment for environmental services, agreements or others.
- The proposal and design of measures for public and recreational use.
- The social and environmental visibility of the Huerta de Valencia.
- The creation of cooperation networks with other territories with similar problems.
- Any other of those foreseen in the objectives of this law and in the territorial action plan that develops it.

The role of this plan is very important and, obviously, as a general idea, must address the agricultural activity to sustainable and ecological way using new technologies and marketing criteria as sales by internet in a global market or guarantee the food sovereignty. All this implies new challenges and the recovery of farmer's role.

3.7 Inventory of professional agricultural holdings

Finally, the law regulates the inventory of professional agricultural holdings as a particular case of the General Registry of Agricultural Production. The L'Horta Council manages this inventory. The objective is to have a list of professional farmers to access, if necessary, processes of leasing abandoned lands or lands whose owners have no interest in cultivating and which are included in the Land Bank. The Land Bank thus becomes an instrument to maintain the land under cultivation that is also managed by the L'Horta Council.

4 CONLUSIONS

The law of L'Horta establishes an innovative regulation for the management of peri-urban agricultural spaces with environmental and landscape values in a metropolitan area. Its contents are the result of a long process of collective maturation of society and of the different disciplines that participate in one way or another in the planning and management of this space.

The strategical actions include a set of strong elements: the development of a spatial planning plan, the creation of a management entity with economic management capacity that has a global vision of the problems and can develop strategic actions in the long term and the development of an agrarian development plan. All this is completed by a series of complementary measures to facilitate the achievement of the objectives of the law.

The law with the territorial plan and the agrarian plan mark the beginning of a new way to understand the territorial management of a metropolitan agricultural land.

ACKNOWLEDGEMENTS

The present work summarises a law that is the result of the work of a large number of stakeholders of the Valencian society. I consider it fair to recognize the continued effort of everyone in general because the result is a collective work of the Valencian people.

In particular, without wanting to be exhaustive and apologizing for persons or organizations not named because of material impossibility, the author wishes to thank the participation of stakeholders who have kept hope alive for many years as the farmer's union *Unió de Llauradors i Ramaders del País Valencià* or the non-governmental organization *Per L'Horta* among many others.

Also to the technicians of the *Conselleria d'Habitatge, Obres Públiques and Vertebració del Territori* of the Autonomous Government of the Valencian Community for their work in elaborate the text of the *Llei de L'Horta* and the *Territorial Action Plan to regulate and revitalize L'Horta de Valencia.* His work has been done through a continuous effort for many years incorporating the results of the various processes of public participation, first by the direction of Mrs. Arancha Muñoz and currently of the General Manager Mr. Lluís Ferrando.

REFERENCES

[1] Cavanilles, A.J., *Observaciones sobre la historia natural, geográfica, agricultura, población y frutos del Reyno de Valencia*, Albatros: Valencia, 1985 (facsimile reproduction 1795 edition).

[2] Miralles i Garcia, J.L., La protección de la Huerta como estrategia de desarrollo económico del entorno metropolitano de Valencia. *Actas Seminario Internacional sobre La Huerta de Valencia*, pp. 359–364, 1994.

[3] Díaz Martín, M. & Galiana Galán, F., *Estudio paisajístico de la Huerta de Valencia*, Ajuntament de València: Valencia, 1996.

[4] Antolín Tomás, C. (coord) et al. *El suelo como recurso natural en la Comunidad Valenciana*, Generalitat Valenciana, Conselleria d'Obres Públiques, Urbanisme i Transports: Valencia, 1998.

[5] Biot Gimeno, C., *La agricultura ecológica, alternativa para la preservación de la huerta de Valencia*. Generalitat Valenciana, Conselleria d'Agricultura, Pesca, Alimentació i Aigua: Valencia, 1998.

[6] The Dobris Assessment Report 1/1995; EUROPEAN UNION. European Environmental Agency Online. https://www.eea.europa.eu/publications/92-826-5409-5 (accessed on 16 May, 2018).

[7] Pla d'Acció Territorial d'Ordenació i Dinamització de l'Horta de València; Conselleria d'Habitatge, Obres Públiques i Vertebració del Territori, Generalitat Valenciana Online. http://www.habitatge.gva.es/ca/web/planificacion-territorial-e-infraestructura-verde/huerta-de-valencia (accessed on 16 May, 2018).

[8] Miralles i Garcia, J.L., Turismo de masas, sostenibilidad y planeamiento. Reflexiones a partir del caso de La Manga (España), *Actas VIII Simposio Universitario Iberoamericano sobre Medio Ambiente, XVII Convención Científica de Ingeniería y Arquitectura*, pp. 149–160, 2014.

[9] García-Ayllón Veintimilla, S., *En los procesos de urbanización del mediterráneo: el caso La Manga*, PhD thesis, Universitat Politècnica de València, València (Spain), 2013.

[10] Miralles, J.L. & García-Ayllón, S., The economic sustainability in urban planning: case La Manga. *WIT Transactions on Ecology and the Environment*, **173**, WIT Press: Southampton and Boston, pp. 279–290, 2013.

[11] García-Ayllón, S. & Miralles, J.L., The environmental impacts of land transformation in the coastal perimeter of the Mar Menor lagoon (Spain). *International Journal of Design & Nature and Ecodynamics*, **9(2)**, pp. 109–128, 2014. https://doi.org/10.2495/dne-v9-n2-109-128

[12] Mateu, A. & Domínguez, M., Cuando El Saler volvió al pueblo; Mètode, 70, 2011 Online. http://metode.cat/es/Revistas/Dossiers/La-Albufera-de-Valencia/Quan-el-Saler-torna-al-poble (accessed on 29 May, 2018).

[13] Miralles i Garcia, J.L., Real estate crisis and sustainability in Spain. *WIT Transactions on Ecology and Environment*, Vol. 150, WIT Press: Southampton, pp. 123–133, 2011.

[14] Muñoz Criado, A., *Plan de la Huerta de Valencia. Un paisaje cultural milenario, Vol. 1 Estrategias de Preservación y Gestión*. Generalitat Valenciana, Conselleria de Medi Ambient, Aigua, Urbanisme i Habitatge: Valencia, 2009.

[15] Muñoz, Criado, A., *Plan de la Huerta de Valencia. Un paisaje cultural milenario, vol. 2 Conclusiones del proceso de participación pública*. Generalitat Valenciana, Conselleria de Medi Ambient, Aigua, Urbanisme i Habitatge: Valencia, 2009.

[16] Romero, J. & Francés, M. (Coord.), *La Huerta de Valencia. Un paisaje cultural con futuro incierto*, Universitat de València: Valencia, 2012.

[17] Guinot, E. & Esquilache, F., La reorganización del paisaje agrario en la Huerta de Valencia después de la conquista cristiana: El sistema hidraúlico y el parcelario de Montcada y Benifaraig en el siglo XIII. *Debates de Arqueología Medieval*, **2**, pp. 229–276, 2012.

[18] Temes, R. & Moya, A., Dynamics of change in the peri-urban landscape of Huerta de Valencia: the case of La Punta (Valencia). *WIT Transactions on Ecology and the Environment,* Vol. 192, WIT Press: Southampton and Boston, pp. 123–131, 2015.

[19] Temes, R. & Moya, A., Typology of the transformations occurred in the peri-urban space of "Huerta de Valencia". Evidence from north arch of Valencia (Spain). *International Journal of Sustainable Development and Planning*, **11(6)**, pp. 996–1003, 2016. https://doi.org/10.2495/sdp-v11-n6-996-1003

[20] Various authors, *Alqueries: Paisatge i arquitectura de L'Horta. València*. Generalitat Valenciana, Valencia, 2004.

[21] Miralles i García, J.L., Díaz Aguirre, S. & Altur Grau, V., Environmental impact on the Mediterranean Spanish coast produced by the latest process of urban development. *WIT Transactions on Ecology and the Environment*, Vol. **155**, pp. 379–389, 2012.

[22] Miralles i García, J.L., Urban coastal development in Valencian country: a paradigmatic case of non-sustainable development. *International Journal of Sustainable Development and Planning*, **10(3)**, pp. 301–314, 2015. https://doi.org/10.2495/sdp-v10-n3-301-314

[23] Gaja i Díaz, F., Hyper planning and the construction bubble on the northern coast of the Valencian countryside (or how the unwarranted scenarios of planning have contributed to the present urban disaster). *WIT Transactions on Ecology and the Environment*, Vol. **192**, pp. 207–217, 2015.

[24] Miralles i García, J.L., The Planning of Peri-Urban Agricultural Areas: the Case of L'Horta de Valencia. *Proceedings of the REAL CORP 2013: Planning Times*, pp. 953–962, 2013.

[25] Miralles i García, J.L., Environmental management of peri-urban natural resources. L'Horta de Valencia case study. *WIT Transactions on Ecology and the Environment*, Vol. 192, pp. 99–110, 2015.

[26] Romero, J. & Melo, C., Spanish Mediterranean Huertas: theory and reality in the planning and management of peri-urban agricultura and cultural landscapes. *WIT Transactions on Ecology and the Environment*, Vol. 193, pp 585–595, 2015.

[27] Garcia, E. & Cabrejas, M., *València, L'Albufera, L'Horta: medi ambient i conflicte social.* Universitat de Valencia: Valencia, 1997.

[28] Marqués Pérez, I., Segura, B. & Maroto, C., Evaluating the functionality of agricultural systems: Social preferences for multifunctional peri-urban agriculture. The "Huerta de Valencia" as case study. *Spanish Journal of Agricultural Research*, **12(4)**, pp. 889–901 2014. https://doi.org/10.5424/sjar/2014124-6061

[29] Propuestas para la transición hacia la gobernanza alimentaria; Ajuntament de Valencia Online. https://valenciacapitalsostenible.org/wp-content/uploads/2017/11/SISTEMATIZACION-SCACS-2017_DEF-2.pdf (accessed on 29 May, 2018).

[30] Miralles i García, J.L., La iniciativa legislativa popular per L'Horta de València. *València, 1808–2015: la historia continua ...* Balandra editions: Valencia, pp. 217–231, 2016.

[31] Ley 5/2018, de 6 de marzo, de la Generalitat, de la Huerta de València. Generalitat Valenciana Online. https://www.dogv.gva.es/datos/2018/03/12/pdf/2018_2459.pdf (accessed on 29 May, 2018).

[32] Water Court Online. http://www.tribunaldelasaguas.org/en/ (accessed on 29 May, 2018).

ARCHITECTURAL IMPLEMENTATION OF VEGETATED COVER FROM AGRICULTURE FOR RESTORING HUMAN THERMAL COMFORT AND MITIGATING THE URBAN HEAT ISLAND EFFECT IN ARID REGIONS

I.E. GAXIOLA[1], N.V. CHALFOUN[2] & C. MOELLER[2]
[1] Arid Lands Resource Sciences, The University of Arizona, USA.
[2] College of Architecture, Planning and Landscape Architecture, The University of Arizona, USA.

ABSTRACT

This investigation describes improved outdoor Human Thermal Comfort levels, based on the effects of integrating vegetated surfaces, such as those from Urban Agriculture systems, to architecture components of a building envelope within Tucson, Arizona, which can contribute on Urban Heat Island mitigation. Urban Agriculture comprises the integration of crop production with the built environment, it can contribute to improving buildings' performance, reducing air pollution, alleviating food scarcity, reducing stormwater runoff, decreasing fossil fuel use, and restoring Human Thermal Comfort. A methodology for outdoor Human Thermal Comfort assessment was applied. It involved the use of digital analysis of fish-eye lens photographs, and 'OUTDOOR', a computer software developed by Nader Chalfoun, Ph.D., at the University of Arizona, which is capable of calculating Human Thermal Comfort indices. Assumptions of this study include: access to water, soil, air, a building envelope, and the presence of vertical and horizontal arrangements of vegetated surfaces, produced in successfully developed Urban Agriculture systems around a selected building envelope in a hot-arid climate. Existing Human Thermal Comfort conditions were compared to those simulated with the integration of vegetated surfaces in order to evaluate the potential effects of Urban Agriculture, and to reach restored Human Thermal Comfort levels.

Keywords: fish-eye lens photograph, hemispherical photography, human thermal comfort, human view-factor, mean radiant temperature, urban agriculture, urban heat island.

1 INTRODUCTION

The built environment is one of the most intricate man-made systems. In view of this, attempting to achieve efficient use of resources and energy while adapting to climate conditions can help determine the physical characterization of a city. From the above, it can be stated that adequate models of human shelter and food production are required to optimally operate while dealing with complex environmental phenomena.

Among these phenomena, Urban Heat Islands (UHI) represent a consequence of intensified human activity, urban sprawl and arguably unwise employment of building materials. It is mainly driven by overconsumption of energy in urban cores and the building materials we use that store more heat compared to natural vegetated conditions. It is commonly characterized as the presence of different levels in temperature when comparing an urban development and its corresponding non-urban surrounding [1] (Fig. 1), and it tends to be stronger during night time [2].

Thermal energy radiated by building materials has a direct impact on Human Thermal Comfort (HTC). It is more difficult for the human body to maintain its optimal temperature level while receiving excessive heat from its surroundings in hot conditions where the air temperature is above the skin temperature and there is no mechanism to dissipate heat [3]. Herein, it is important to take careful consideration of the geometry of space [2], and thermal properties of materials (e.g. albedo, emissivity, specific heat, heat capacity, and surface

DOI: 10.2495/DNE-V13-N4-373-383

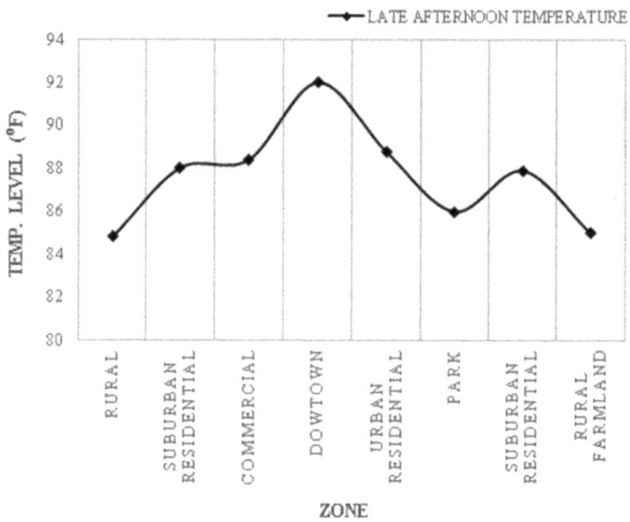

Figure 1: Sketch of urban heat island effect (Reproduced from http://ei.lehigh.edu/eli/luc/resources/uhi_profile.gif).

temperature), particularly in outdoor spaces. Smart use of building materials and, the implementation of green infrastructure can contribute to decreasing the impact of UHI [4].

Different green infrastructure systems which can be incorporated to the design of buildings, for example, green roofs, porous pavement, rain gardens, bioswales, planting trees and shrubs, etc. [5]. As a special mechanism for growing crops within the built environment [6], Urban Agriculture (UA) can be considered as an additional category within green infrastructure approaches. Different UA systems can be adopted, according to spatial conditions of the built environment, to effectively produce urban vegetation. For example, roof gardens, hydroponics, aquaponics, greenhouses, plant factories with artificial light, and remote skylighting.

Roof gardens involve the growing of plants on rooftops and, are very useful for rainwater harvesting. Hydroponics involve a recirculating water system to provide nutrients for plants and do not require soil: an inert mean for the plant to grow is provided. Aquaponics are based on hydroponics, with the difference that in aquaponics, plant nutrients are supplied by fish-generated waste, which is chemically processed by bacteria that plants metabolize to grow. Greenhouses comprise a confined surface area covered with a transparent-translucent material, operate using multiple systems (ventilation, refrigeration, irrigation, lighting, fertigation, etc.), and can be energy efficient. However, to build and operate an energy-efficient greenhouse requires a certain level of technical knowledge on controlled environment systems [7].

A plant factory with artificial light is an indoors-controlled environment system that supplies artificial light for plants to grow [8]. Remote skylighting combines fiber optics and a geometric mechanism to collect sunlight and channel it to a desired point using a concave dish oriented towards the sun, and a distributor dish on the other end of the system [9].

From the above, it can be stated that successful UA systems require access to solar radiation, space to grow (e.g. land, building components, such as facades), infrastructure, soil, water, institutional support, and participatory citizens [10], [11]. Its contribution lies partly on the fact that cities lack the means to become self-sufficient regarding food production thus, UA implies a reduction in energy use on transportation of food, and an increase in the amount of vegetated cover [5], which could improve thermal conditions of the built environment. Increasing evapotranspiration through vegetated surfaces in the built environment contributes to decreasing latent heat, otherwise, a greater amount of energy ends up being available as sensible heat, which represent an increase in surface temperature levels and potentially a more intense UHI [12]. For example, approximately half of a building cooling loads can be reduced, if its thermal performance is mostly defined by the building envelope, and it is appropriately shaded using vegetation. The effect will be less for internal load dominated buildings, such as typical large offices, data centers, factories, or hospitals. For this purpose, plant species need to be properly selected and located considering climate and building shape [13].

Appropriate combination of water components and vegetation contributes to decreasing temperature levels in outdoor spaces [14]. Temperature levels of bodies of water tend to be lower compared with surrounding building materials, to the point of accounting for a difference between 2 and 6°C [12]. Additional benefits include the option to store rainwater for irrigating UA and contribute to sequestering stormwater runoff [5].

Coupling aquaculture and agriculture outdoors is an example of how UA systems can aid on cooling outdoor spaces, not only by transpiration and shade provided by plants, but also by using water for direct evaporative cooling thus, the effective combination of water features and urban agriculture can contribute on mitigating UHIs [12]. Assumptions considered for this study were: (1) Proposed vegetated cover was regarded as a product of successfully developed UA systems in a hot arid climate, (2) Access to water, soil, air, and a building envelope was provided, (3) For analysis purposes, proposed water features were considered as always filled with water.

2 OVERVIEW OF METHODOLOGY

Hemispherical Photograph Analysis (HPA) was applied to five selected locations at Inna E. Gittings building, located within the main campus of the University of Arizona in Tucson, AZ. The execution of HPA, based on Watson and Johnson's method [15], allowed to estimate human thermal comfort indices in outdoor spaces.

First, site locations are selected according to study-specific thermal circumstances. Subsequently, the percentage of Human View Factor (HVF) of materials affecting HTC is calculated for each location. HVF is useful to define the portion of radiant energy received by a person from a source surface, in relation to a particular location in space [16]. This step involves the use of two fish-eye lens photographs with a field of view of 180 degrees each. One photograph is taken aiming to the sky, and the second one pointing to the ground, in order to account for a sphere sample. The camera must be levelled, oriented to the north, and elevated 3 feet from the ground, if the analysis considers a person in standing position (Fig. 2). Photograph samples are scaled to fit a 6-inch diameter circumference. Samples are overlapped with a 6-inch diameter circular grid divided into 1000 equal units of 0.02827 in^2 each. The grid represents a polar diagram of HVF for an individual in standing position [15] (see Fig. 2).

Processing of samples, for accounting surface area of building materials, can be manual or automatic. The former requires printing the polar grid in a translucent material and overlapping it with the fish-eye photograph. Then, the amount of grid units per each material can be accounted and then multiplied by 0.02827. The automatic method is more accurate and less time consuming, it involves the use of computer software to filter and trace the perimeters or shapes of each defined material and, as a result of the process, surface areas of different materials can be automatically estimated.

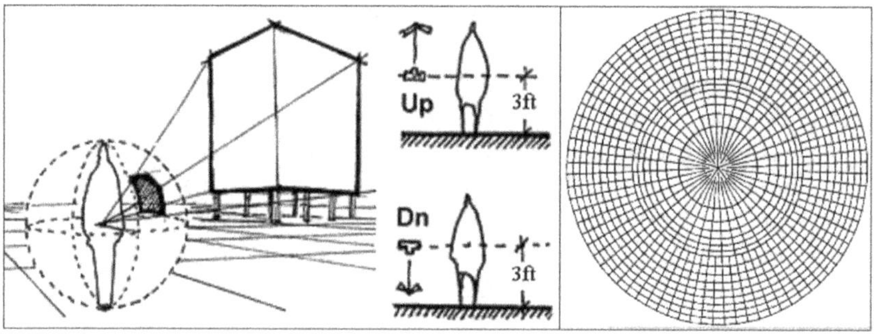

Figure 2: Scheme of radiant field affecting an individual. Left: representation of individual's view-factor and camera orientations for shooting fish-eye lens photographs *(Chalfoun, 2002)*. Right: polar diagram of human view-factor for an individual in standing position (*Watson and Johnson, 1988*).

When complex shapes of vegetation show in photographs, effective extraction of surface area requires the combined use of vector- and pixel-based software. After estimating the surface area of each material within each sample, HVF is calculated using eqn (1).

$$HVF = \left(\frac{SAM}{GUA}\right)0.0005 \tag{1}$$

Where:

- HVF = Human View Factor;
- SAM = Surface Area of Material (in^2) considering both hemispheres;
- GUA = Grid unit area, taken as 0.02827 in^2. It is a unit from a thousand in which a circular area with a diameter of 6 inches is divided, see right side of Fig. 2;
- 0.0005 = Correction factor of unity as a conservation of radiation leaving all surfaces that affect the two hemispheres of the human comfort spherical region.

Once the HVF has been calculated, meteorological and geographical data, as well as thermal properties of materials can be supplied to 'OUTDOOR' software. The data can be gathered on site, extracted from a Typical Meteorological Year (TMY) file, and from existing literature or official websites. An on-site survey of surface temperature of materials can be employed to provide a detailed profile of surface temperature variation throughout the day. Noncontact temperature measurements of close range targets can be made using an infrared thermometer.

An adequate temporal resolution aids on detecting trends in thermal comfort indices; hence, before employing 'OUTDOOR', a total amount of tests per day requires to be specified. For this study, six (one every four hours) were the selected amount of calculations. After estimating human thermal comfort indices in 'OUTDOOR', worst-case scenarios are identified then, improved spatial conditions can be proposed using a digital 3D model. Finally, fish-eye lens images generated from the digital 3D model are processed, and 'OUTDOOR' estimations are performed. This iterative process goes on until satisfactory HTC levels are reached.

3 HEMISPHERICAL PHOTOGRAPH ANALYSIS AND 'OUTDOOR' SOFTWARE

What follows are five steps describing HPA application to fish-eye photograph samples gathered at Ina E. Gittings Building, located within the Main Campus of the University of Arizona, in the hot-arid city of Tucson, AZ; and the use of 'OUTDOORS' for estimating HTC indices.

3.1 Site identification

Five outdoors locations were selected from Inna E. Gittings building, in the north-east region of the University of Arizona main campus (Fig. 3, center). Since the building dates from 1964, structural deficiencies could be a concern if heavy UA systems were to be attached to it. However, what made it suitable as a research case for an UA study were its large and simple walls, its strategic location within campus, presence of vegetation, and the access to water granted by existing infrastructure.

Criteria to select sampling locations consisted on identifying presence of vegetated surfaces, prevailing materials, geometry of space (particularly openness to the sky), and potential to integrate infrastructure of UA systems.

Location 1 is an uncovered esplanade with presence of vegetation on the north side, and neighboring an asphalt street on the south. Location 2 is an uncovered walkway with presence of vegetation, dirt, and high emissive materials. Location 3 is an uncovered courtyard with scarce vegetation and high emissive materials. Location 4 is a narrow, uncovered hall with no vegetation, flanked by high brick walls. Location 5 is a roof located on the south-west zone of the site. It has high reflectance and no vegetation (Fig. 3, center).

3.2 Gathering fish-eye lens photographs

Useful photographs were captured when uniform lighting conditions were present on building materials, particularly a few minutes after sunset [17]. A Nikon D610 Fx DSLR camera, with a SIGMA 8mm F4 circular image fish-eye lens attached, was mounted on a tripod, oriented to magnetic north, levelled, and the tip of the lens was set to a height of 3 feet for both, sky and ground samples (Fig. 4, left). Examples of useful photographs are: sample 02 (F-stop = f/16, exposure = 1/80 sec, and ISO = 6400), sample 03 (F-stop = f/11, exposure = 1/400 sec, and ISO = 6400), and sample 04 (F-stop = f/8, exposure = 1/1600 sec, and ISO = 6400), see right side of Fig. 4.

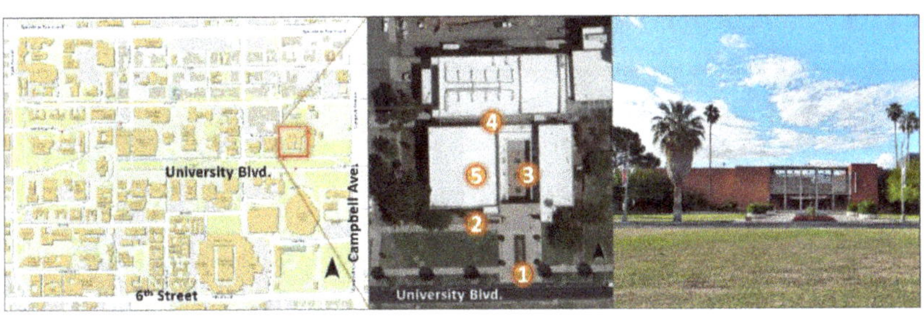

Figure 3: Study site. Left: aerial view of U of A main campus (*GIS MapTucson and U of A map*). Center: plan view of selected locations (*Google earth*). Right: south façade of case study building (*Gaxiola, 2016*).

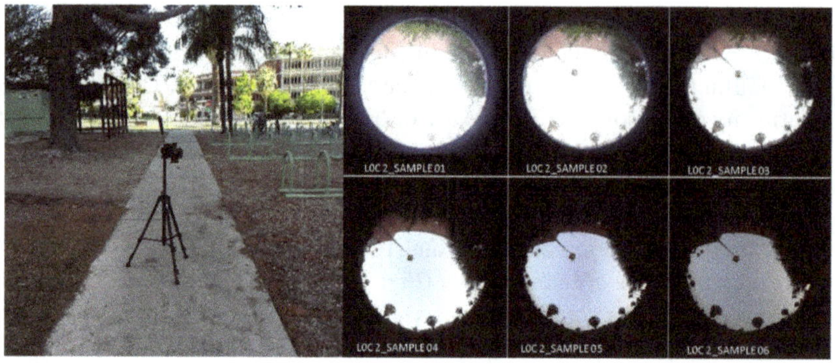

Figure 4: Fish-eye lens photograph samples. Left: camera set for a ground sample. Right: six samples aiming to the sky at location 2.

3.3 Digital analysis of fish-eye lens pictures

This phase was developed using AutoCAD, Photoshop, ImageJ, and the Photo editor from Microsoft. Two distinct cases were found while analyzing fish-eye pictures. The first had scarce or no vegetation shown in the pictures, the second case had abundant vegetation overlapping with surfaces of building materials.

The first step for accounting surface area was to scale each fish-eye picture in AutoCAD so that the useful circular area had a 6-inch diameter. After visually identifying materials, vectors were traced on their perimeters (Fig. 5, center). Then, vectors were set to layers, solid-color regions were generated and, if no vegetation was present, surface area was directly estimated in AutoCAD (Fig. 6, right).

For estimating surface area in pictures with abundant vegetation, the layer including all plant species was filtered using Photoshop, and then it was subtracted from the building material of interest and, after adding a 1-in^2 black square as scale reference, surface area was retrieved using particle analysis in 'ImageJ' (Fig. 7).

Once the surface area was calculated for each material in both fish-eye pictures of every location, HVF was calculated using eqn (1). Resulting HVF for each individual material was fed in 'OUTDOOR', in addition to meteorological and geographical data, adequate Clothing Insulation factor (CLO), Metabolic Rate values (MET), and thermal properties of materials gathered on site. For this study, a value of 0.5 for CLO, and 2.0 for MET were considered as applicable for users performing academic activities during summer in the main campus of the

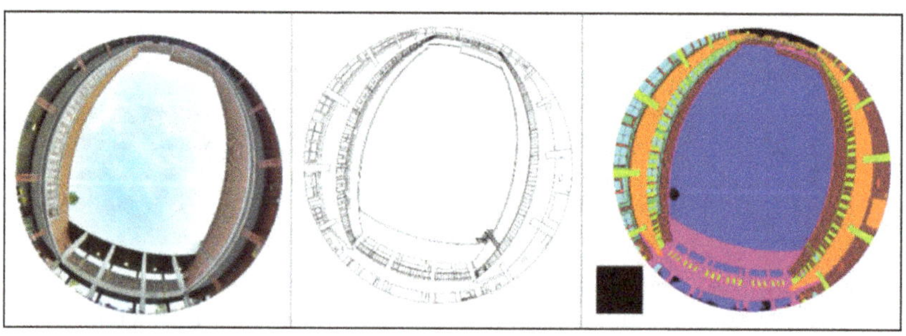

Figure 5: Sky photograph sample, location 3. Left: original photograph. Center: vectored sample. Right: Materials layered in color, the black square is the scale reference.

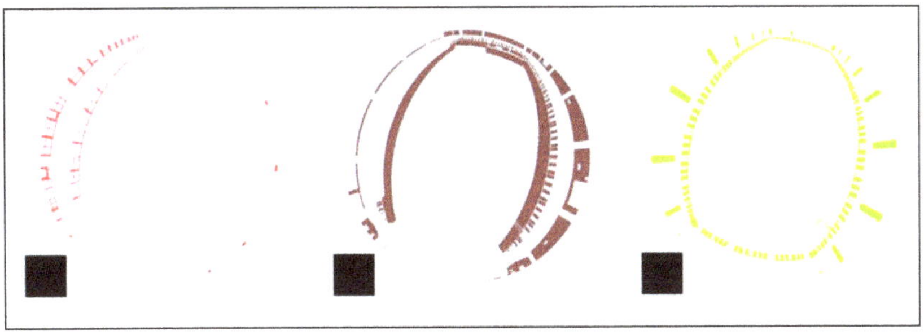

Figure 6: Three vectorised materials from location 03 after being processed in AutoCAD. Left: Aluminium. Center: Brick. Right: Iron.

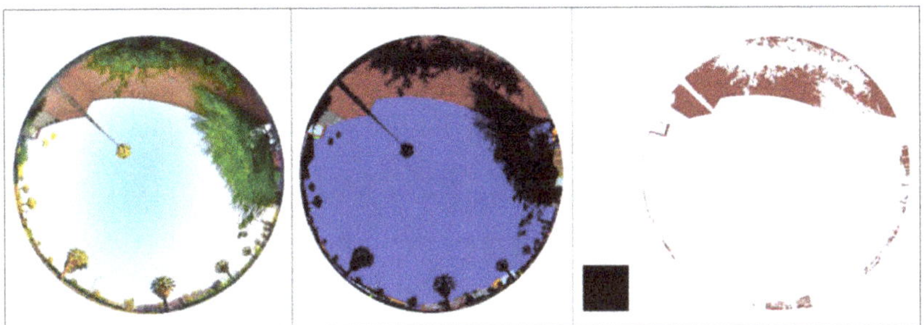

Figure 7: Sky fish-eye lens photograph, location 2. Left: original photograph. Center: vectorised materials. Right: final version of a sample material (brick) ready to be analysed in 'ImageJ'.

University of Arizona, in the hot-arid city of Tucson, AZ. Results from 'OUTDOOR' analyses were employed to detect trends in thermal conditions of the space in question, and worst case scenarios were identified.

3.4 Data Interpretation, balancing levels of thermal radiation

After performing 'OUTDOOR' estimations, location 03 (L03) was selected as the worst case scenario and; as suitable for UA implementation. Initial HVF on L03 showed highly radiating surfaces (Fig. 8). Considering a Predicted Mean Vote (PMV) range from −0.5 to +0.5 as comfortable [18], 'OUTDOOR' estimations on L03 at 8:00am, 12:00pm and, 4:00pm with 0% of shade, presented the highest uncomfortable PMV levels: 2.17, 2.75 and, 3.18, respectively.

'OUTDOOR' results aided on identifying dominant conditions of high PMV values: overexposure to radiation from brick, concrete, and open sky. Then, an alternative balanced HVF (Fig. 8) was proposed and used in 'OUTDOOR' with the scope of reaching a restored thermal comfort condition.

3.5 Digital modelling HTC restoration

After examining HTC indices using 'OUTDOOR', a digital 3D model of the building envelope was created using SketchUp, including landscape, urban furniture, and surrounding buildings (Fig. 9). Highly accurate fish-eye lens images were developed from the digital 3D

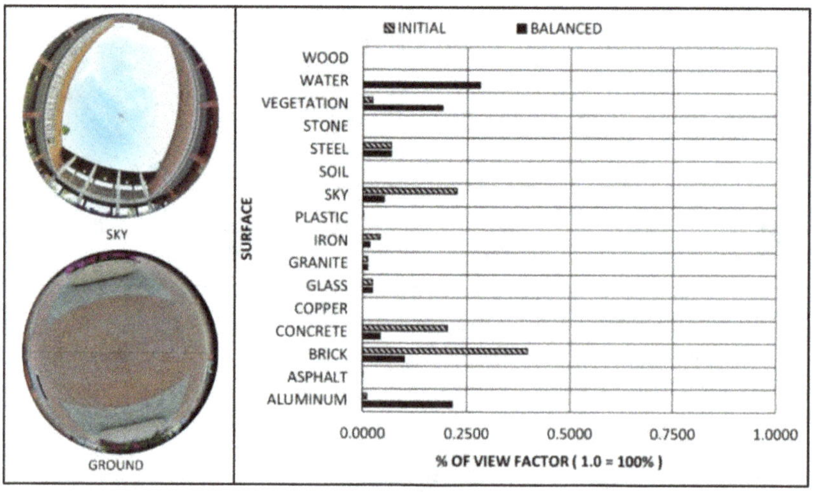

Figure 8. Left: fish-eye photographs of Location 03. Right: %VF values before and after balancing radiating surfaces.

Figure 9: Existing conditions of the site. Digital 3D model created in SketchUp. Left: Southwest aerial perspective. Right: Northeast aerial perspective.

model using a rendering engine called POV-Ray: a fish-eye lens camera was configured and located in the digital 3D model on the same position as in the actual site (Fig. 10, center). Then, HVF was directly measured in 'ImageJ' using fish-eye rendered samples of isolated materials (see right side of Fig. 10).

This method allowed to execute faster 'OUTDOOR' estimations of environmentally retrofitted scenarios created in the digital 3D model. According to a climate analysis for the city of Tucson executed in Climate Consultant, in order to grant outdoor thermal comfort during critical hours (June 21st at 4 pm on Location 3 of the research case), selected strategies were: shading, vegetated cover, and water bodies. The combination of these strategies can improve HTC mainly through direct evaporative cooling, evapotranspiration, and shade [14], while also allowing to potentially regulate incident sunlight access for potential UA development.

In order to find the suitable HVF of every material, an iterative process was executed from 'OUTDOOR' back to the digital 3D model, and vice versa. Improvements were achieved by decreasing surface area of brick and concrete, adding vegetation, increasing shade, and incorporating a water body. The location and geometry of the spatial renovation was designed considering potential UA agriculture systems. As shown in Fig. 11, horizontal and vertical

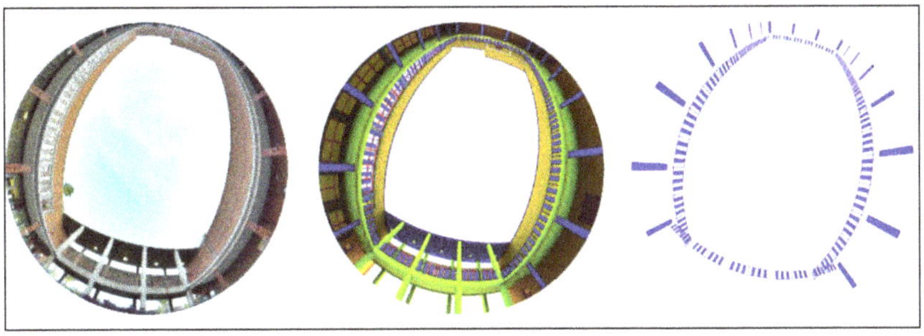

Figure 10: Comparison between a real fish-eye lens sample (left) and an image rendered in POV-Ray using a fish-eye camera on the same location (center). Fish-eye sample of a single material (right).

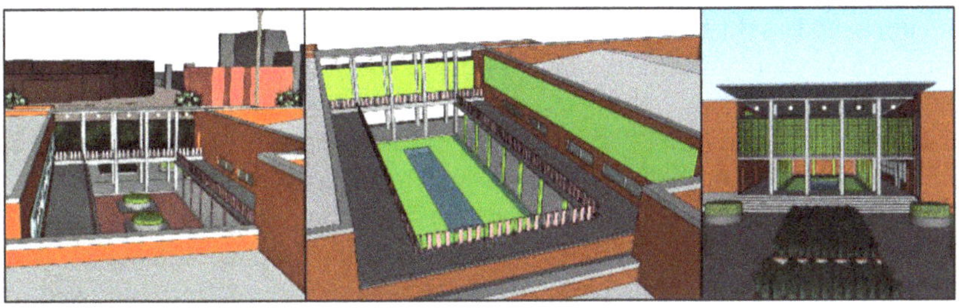

Figure 11: Perspectives of proposed improved conditions. Left: original building conditions. Center: Aerial view of courtyard with vegetated surfaces and central body of water. Right: south façade showing a horizontal aluminium shading structure.

planes in green represent vegetated surfaces, a body of water was proposed in the middle of the courtyard, and a louvered horizontal structure made out of aluminium was proposed on top of Location 03, in order to control solar radiation.

Noticeable improvements in HTC were found after implementing green infrastructure, considering the option of 100% shading conditions. At selected times when the building in question is operating (8:00, 12:00 and 16:00 hours), PMV levels decreased from 2.17, 2.75 and 3.18 to complying levels of -0.4, 0.5 and 0.5, respectively [18] (Fig. 12). MRT levels

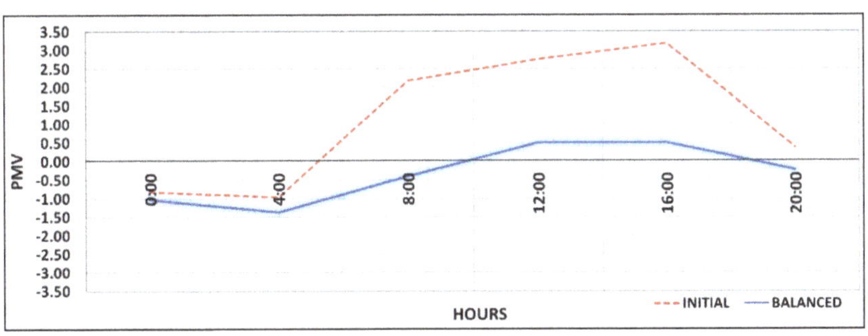

Figure 12: Line chart showing decrease on PMV levels after applying green infrastructure to the building envelope. The red dotted line represents PMV results in conditions with no shade. The blue line belongs to PMV values of improved conditions with full shade.

were reduced, for the hours mentioned above, from 130.4°F, 149.4°F and 170.8°F to 40.78°F, 60.41°F and 58.62°F, respectively. Effective Temperature (ET) levels of the final design proposal were reduced to 64.66°F, 74.80°F and, 74.72°F for 8:00, 12:00, and 16:00 hours, respectively.

4 CONCLUSIONS

Digital processing and generation of fish-eye lens photographs, aided by using 'OUTDOOR' software, allowed to analyze, simulate, and propose outdoor HTC conditions of an educational building located in the hot-arid city of Tucson, AZ. Simulations of a retrofitted outdoor space proposal, which incorporated strategic location of vegetated cover, a water feature, and shade, showed comfortable levels of PMV. By creating fish-eye lens images using a rendering engine, it can be stated that on site fish-eye samples are not completely necessary.

At building scale, the location and amount of surface materials is directly related to HTC. For example, the potential thermal impact/benefit of a surface material can increase as it is closer to human bodies. In addition to outdoor spaces being suitable for implementing urban agriculture, indoors implementation can be of use to reach better air quality, and psychological comfort.

Water bodies can be incorporated, and coupled with vegetated cover outdoors as direct evaporative cooling components, and as an irrigation resource for UA.

Thermal conditions in outdoor spaces can be optimized by regulating short- and long-wave radiation to provide HTC, with the additional benefit of crop production. Thus, excessive heat can be managed at urban scale by replicating and strategically placing improved microclimate outdoor thermal scenarios that include UA.

Future research avenues include determining a total amount of vegetated cover for decreasing the UHI effect in a given city, the use of drones for gathering fish-eye picture samples, and studying the thermal effect of individual plant species.

REFERENCES

[1] Cermak, J.E., Davenport, A.G., Plate, E.J. & Viegas, D.X. (eds), *Wind Climate in Cities* [Internet]. Springer Netherlands: Dordrecht, 1995 [cited 31 December, 2015]. Available from http://link.springer.com/10.1007/978-94-017-3686-2

[2] Oke, T.R., The energetic basis of the urban heat island. *Quarterly Journal of the Royal Meteorological Society*, **108(455)**, pp. 1–24, 1982.

[3] Moe, K., *Thermally Active Surfaces in Architecture*, p. 240, 1st edn., Princeton Architectural Press: New York, 2010.

[4] US EPA. Heat Island Cooling Strategies [Internet]. 2015 [cited 17 May, 2016]. Available from https://www.epa.gov/heat-islands/heat-island-cooling-strategies

[5] Daniels T. *The Environmental Planning Handbook for Sustainable Communities and Regions*, p. 548, 2nd edn., APA Planners Press: Chicago, 2014.

[6] Lupia, F. & Pulighe, G., Water use and urban agriculture: Estimation and water saving scenarios for residential kitchen gardens. *Agriculture and Agricultural Science Procedia*, **4**, pp. 50–58, 2015. https://doi.org/10.1016/j.aaspro.2015.03.007

[7] Philips, A., *Designing Urban Agriculture: A Complete Guide to the Planning, Design, Construction, Maintenance and Management of Edible Landscapes*, p. 288, 1st edn., Wiley: Hoboken, New Jersey, 2013.

[8] Toyoki Kozai. What PFAL means to urban agriculture. *Resource Magazine*, **20(2)**, p. 12, 2013.

[9] Drumm, P., The LowLine: How Do the Remote Skylights Work? [Internet]. Core77, 2012 [cited 5 May, 2016]. Available from: http://www.core77.com//posts/22088/The-LowLine-How-Do-the-Remote-Skylights-Work

[10] Jacobi, P., Drescher, A. & Amend, J., Urban Agriculture Justification and Planning Guidelines. Urban Vegetable Promotion Project [Internet], 2000 May [cited 9 April, 2018]; Available from https://d3n8a8pro7vhmx.cloudfront.net/erc/pages/46/attachments/original/1447203906/Urban_Agriculture_-_Justification_and_Planning_Guidelines.pdf?1447203906

[11] Angotti, T., Urban agriculture: long-term strategy or impossible dream?: Lessons from Prospect Farm in Brooklyn, New York. *Public Health*, **129(4)**, pp. 336–341, 1 April, 2015. https://doi.org/10.1016/j.puhe.2014.12.008

[12] Qiu, G., Li, H., Zhang, Q., Chen, W., Liang, X. & Li, X., Effects of evapotranspiration on mitigation of urban temperature by vegetation and urban agriculture. *Journal of Integrative Agriculture*, **12(8)**, pp. 1307–1315, 1 August, 2013. https://doi.org/10.1016/s2095-3119(13)60543-2

[13] Lechner, N., *Heating, Cooling, Lighting: Sustainable Design Methods for Architects*, p. 912, 4th edn., Wiley: Hoboken, New Jersey, 2015.

[14] Robitu, M., Musy, M., Inard, C. & Groleau, D., Modeling the influence of vegetation and water pond on urban microclimate. *Solar Energy*, **80(4)**, pp. 435–447, 1 April, 2006. https://doi.org/10.1016/j.solener.2005.06.015

[15] Watson, I.D. & Johnson, G.T., Estimating person view-factors from fish-eye lens photographs. *International Journal of Biometeorology*, **32(2)**, pp. 123–128, June 1988. https://doi.org/10.1007/bf01044905

[16] Chalfoun, N.V., Sustainable urban design and outdoor space analysis using MRT, and photography of scale models: A case study of the Rio Nuevo project in Tucson, Arizona, USA. In *PLEA* [Internet]. Toulouse, France, 2002 [cited 22 May, 2016]. Available from http://capla.arizona.edu/hed/docs/PleaToulouse.pdf

[17] Rich, P., *A Manual for Analysis of Hemispherical Canopy Photography. Los Alamos National Laboratory Report*. 1 January, 1989, LA-11733-M.

[18] Anderson, K., *Design Energy Simulation for Architects: Guide to 3D Graphics* [Internet], p. 272, Routledge: New York, Oxfordshire, England, 2014. Available from: https://www.amazon.com/Design-Energy-Simulation-Architects-Graphics/dp/041584066X/179-9162303-0525732?ie=UTF8&*Version*=1&*entries*=0

EFFECTS OF VERTICAL GREEN TECHNOLOGY ON BUILDING SURFACE TEMPERATURE

I. BLANCO[1], E. SCHETTINI[2] & G. VOX[3]
Department of Agricultural and Environmental Science DISAAT – University of Bari, Italy.

ABSTRACT
A sustainable technology for improving the energy efficiency of buildings is the use of urban greening in order to reduce the energy consumption for air conditioning in summer and to increase the thermal insulation in winter. A worldwide growing interest in urban green is encouraging the application of the greening technology for more sustainable buildings. Building indoor air temperature depends on several different parameters related to the climate of the region, the building itself and its use. The main parameters influencing the microclimate are: external air temperature and relative humidity, incident solar radiation, long wave radiation exchange between the building surfaces and its surroundings, wind velocity and direction, air exchanges, physical and thermal properties of the building's envelope materials, design variables such as building dimensions and orientation, presence of artificial light, electrical equipment. Green façades can allow the physical shading of the building and promote evapotranspiration in summer and increase the thermal insulation in winter. External wall surface temperature is a parameter useful to assess the effectiveness of green façades. An experimental test was carried out at the University of Bari (Italy) for three years. Three vertical walls, made with perforated bricks, were tested: two were covered with evergreen plants (*Pandorea jasminoides variegated* and *Rhyncospermum jasminoides*) while the third wall was kept uncovered and used as control. Several climatic parameters concerning the walls and the ambient conditions were collected during the experimental test. The experimental data were used for developing a multiple regression equation regarding the dependence of the difference of external surface temperature between the green façades and the control wall and the weather conditions. The model shows a good predicting ability.
Keywords: air-conditioning, energy savings, green walls, regression model, urban agriculture, urban heat island.

1 INTRODUCTION

Urban Green Infrastructures (UGI) contribute to improving urban climate and reducing urban air temperature, maximum surface temperature and surface temperature variation especially in regions characterized by Mediterranean climate [1]–[5].

A worldwide growing interest in urban greenery is stimulating the application of the greening technology for more sustainable buildings [6]–[8]. The urban greening allows to improve the energy efficiency of buildings by reducing the energy consumption for air conditioning in summer and by increasing the thermal insulation in winter [9]–[15]. Building indoor air temperature depends on several different parameters related to the climate of the region, the building itself and its use. The main parameters influencing the microclimate are: external air temperature and relative humidity, incident solar radiation, long wave radiation exchange between the building surfaces and its surroundings, incidence and speed of the wind, air exchanges, physical and thermal properties of the building's envelope materials, design variables, such as building dimensions and orientation, presence of artificial light, electrical equipment.

[1] *ORCID: http://orcid.org/0000-0003-2927-3427*
[2] *ORCID: http://orcid.org/0000-0002-8456-8677*
[3] *ORCID: http://orcid.org/0000-0003-4017-5174*

DOI: 10.2495/DNE-V13-N4-384-394

UGI include engineered options such as vertical greenery systems in a building that are a rising strategy in high-density urban development with scarce green areas and limited open ground space [1], [15]–[17].

Green façades and living walls are vertical greenery systems [18]–[19]. The green façade is characterized by the growing media staying on the ground at the base of the building or at different heights of the façade and by plants growing vertically in order to cover the vertical building surfaces [13]–[20]. The presence or absence of a support leads to a further distinction in direct and indirect façade. In the direct green façade (or traditional green façade), plants climb directly on the façade of the building due to their morphological features. In the indirect green façade (or double-skin green façade), plants are supported by a structure such as mesh, wire, trellis, which can be located on a free-standing system, such as fence or columns or even to a short distance from the wall [16]–[21]. The living wall has the growing media standing vertically in front of the vertical surface [13]. The living wall can be fixed to a wall or to a free-standing frame [17], [20], [22]. The living wall has major aesthetical potentialities due to the possibility of installing a wider range of plants in comparison to a green façade. The living wall is a more elaborate and costly technology, also with respect to the maintenance and energy costs [12].

Greenery vertical systems moderate the indoor thermal environment by handling the building envelope heat transfer processes (radiative, convective and conductive) [23]–[25]. The transfer of a heat wave through a vertical green is a multifaceted phenomenon due to the interposition of the plant layer between the outdoor environment and the building envelope. Green vertical systems can permit the physical shading of the building and promote evapotranspiration in summer and increase the thermal insulation in winter [26]. The external layers of a green cover behave like optical filters; the profounder layers behave like an insulation material. The leaves of the plants create a quite steady layer of air, acting as an additional thermal insulation layer that lowers the wind strength. The solar absorption coefficient value for the vertical green has been estimated to be about one third of that for an exposed bare wall. The insulating behaviour (thermal resistance) of the plant layer becomes dominant when the density of plant canopy, i.e. the covering ratio, reaches 100% [23].

Green wall thermal performance should be analysed for identifying which systems are the most effective in bringing energy savings for building cooling/heating. Energy savings heavily depend on a multitude of factors. The thermal performance of the vertical green systems in summer needs further study with reference to the relationship with the climatic conditions and to the plants that can be actually useful in a given context, in order to fill the gaps in literature. External wall surface temperature is a parameter useful to assess the effectiveness of green vertical system. It is also the most commonly observed parameter on the building or experimental prototype. Despite its evaluation does not allow adequately quantifying microclimatic benefit and thermal performance [27], it permits to compare alternative designs related to the adoption of different plant species used, or the distance of the plant layer from the wall, or the typology of building envelope.

The analysis of alternative system configurations can be evaluated using simulations in order to determine the most efficient options. Simulations can be made by different types of approaches, such as the statistical, the hybrid and the engineered ones [28]. The statistical technique called Multiple Linear Regression needs large amount of experimental data. It is ease to develop and to interpret for forecasting, in comparison to other techniques [29]. Simulation engineered models often were not validated with experimental data [2], [17], [27].

This research aims to analyse the thermal performance of two different green façades in the summer period in the Mediterranean region. An experimental test was carried out on two small-scale green façades at the experimental field of the University of Bari, South Italy. The effects of the greenery systems on the external surface temperature of the walls were reported. Moreover, the relationship between the external climatic conditions and the external surface temperature has been modelled on 2015 data through a regression technique. The model was validated on 2016 data. It can be used for predicting the external surface temperature reduction by the green cover.

2 MATERIALS AND METHODS

The experimental field of the University of Bari in Valenzano (Bari, Italy), at latitude 41° 05' N, longitude 16° 53' E, altitude 85 m ASL, is characterized by a Mediterranean climate.

Three identical wall prototypes were built taking into consideration a typical Mediterranean building solution, i.e. a single skin of perforated bricks joined with mortar (Fig. 1). Each wall, south oriented, measured 1.00 m in width, with a height of 1.55 m. The walls were 0.22 m thick. The masonry and the plaster coating were 0.20 m and 0.02 m thick, respectively. The average density of the walls was 695 kg m^{-3} (plaster included). The thermal conductivity coefficients of the masonry and the plaster coating were 0.28 $Wm^{-1}K^{-1}$ and 0.55 $Wm^{-1}K^{-1}$, respectively [28]. The heat capacity coefficient was 840 $Jkg^{-1}K^{-1}$ for the masonry and 1000 $Jkg^{-1}K^{-1}$ for the plaster coating.

The walls were shielded and insulated on the backside with a structure made of sheets of expanded polystyrene (Fig. 1). Moreover these polystyrene structures were externally covered with a coloured net in order to reduce the effect of the incident solar radiation.

Figure 1: The experimental green façades at the University of Bari; the left wall is the reference uncovered wall, the central wall is covered with *Pandorea jasminoides variegated* and the right wall with *Rhyncospermum jasminoides*.

One wall without plants was the reference. The other walls were covered on June 18, 2014 with an evergreen climbing plant (one with *Pandorea jasminoides variegated,* the other *Rhyncospermum jasminoides*). The plants were chosen for their suitability for Mediterranean climate. The coverage of the walls has gradually increased, starting from June 2014. An iron net was placed at a distance of 15 cm from the walls for supporting the plants vertical growing. The plants were irrigated with the drip system and fertilized with an N:P:K 12:12:12 fertilizer. The plants were characterized by a Plant Leaf Area Index (LAI) ranging from 2.0 to 4.0 for *Rhyncospermum jasminoides* and from 1.5 to 3.5 for *Pandorea jasminoides variegated* throughout the year. LAI was measured with a AccuPAR PAR/LAI Ceptometer (model LP-80, Decagon Devices Inc., Pullman, WA, USA) once plants were fully developed.

Several climatic parameters concerning the ambient conditions and the walls behind the plants were collected at the experimental field from June 2014. These parameters were: the temperature of the plaster on the external surface of the walls, the external air temperature, the solar radiation on a horizontal and on a vertical plane, the relative humidity of the external air, the wind velocity and direction.

Data were measured with a frequency of 60 s; afterward they were averaged every 15 min and stored by means of a data logger (CR10X, Campbell, Logan, USA). The temperature of the external plaster surfaces was measured by thermistors (Tecno.EL s.r.l. Formello, Rome, Italy). Hygroclip-S3 sensors (Rotronic, Zurich, Switzerland) measured the temperature and the relative humidity of the external air. A pyranometer (model 8-48, Eppley Laboratory, Newport, RI, USA) was used for measuring both the solar radiation incident on a horizontal and on a vertical plane in the wavelength range 0.3–3.0 mm. The wind velocity and direction were measured by means of a Young Wind Sentry anemometer (Young Company, Traverse City, MI, USA). The sensors were shielded from solar radiation.

The analysis of variance (ANOVA) was developed with the CoStat software (CoHort Software, Monterey, CA, USA). It was aimed to identify significant differences between the temperature data on the walls.

The multiple linear regression technique was used to define the linear dependence of the difference of the external surface temperature between the control wall and the green façade, the dependent variable, on the independent parameters, known as 'predictors' [29]–[30].

The regression models can diversify with the kind of data (field data or simulated data), the data time interval and the forecasting temporal horizon [30]. Consequently, the experimental data were grouped into time slots having a similar solar radiation on a horizontal surface (I_{hor}).

The external climate parameters, as external air temperature and relative humidity, horizontal and vertical solar radiation, wind velocity and direction, are the predictors. The difference of the external surface temperature between the control wall and the green façade (Y_t) was expressed by the following equation:

$$\begin{aligned} Y_t &= \beta_0 + \beta_1 Y_{t-1} + \beta_2 X_{1,t} + \beta_3 X_{1,t-1} + \beta_4 X_{2,t} + \beta_5 X_{2,t-1} + \beta_6 X_{3,t} + \beta_7 X_{3,t-1} \\ &+ \beta_8 X_{4,t} + \beta_9 X_{4,t-1} + \beta_{10} X_{5,t} + \beta_{11} X_{5,t-1} + \beta_{12} X_{6,t} + \beta_{13} X_{6,t-1} + \varepsilon_t . \end{aligned} \tag{1}$$

In eqn (1) Y_t and Y_{t-1} are the difference of the external surface temperature at time t and t-1, where t stands for the quarters of hours falling within the period considered. $X_{i,t}$ and $X_{i,\,t-1}$ stand for the predictor variables at time t and at time t-1, respectively: $X_{1,\,t}$ and $X_{1,\,t-1}$ are the external air temperature, $X_{2,\,t}$ and $X_{2,\,t-1}$ are the solar radiation on a horizontal surface, $X_{3,\,t}$ and $X_{3,\,t-1}$ are the relative humidity of the external air, $X_{4,\,t}$ and $X_{4,\,t-1}$ are the wind velocity, $X_{5,\,t}$

and $X_{5,\ t-1}$ are the wind direction, and $X_{6,\ t}$ and $X_{6,\ t-1}$ are the solar radiation on a vertical surface. β_i are the coefficients of the regression model, $i = 0, \ldots, 13$. β_0 is the intercept and ε_t is an error value that estimates the difference between the observed data and the predicted data.

Equation (1) was applied for predicting the temperature difference values. The fitted/predicted value and the parameters of the estimated regression model where determined with the Least Squares Method using the Regression Tool in Excel's Data Analysis add-in. The multiple linear regression model was fitted on the data gathered during August 2015. The model was used for predicting the thermal performance of the green façades during August 2016.

3 RESULTS AND DISCUSSION

During August 2015 the external air temperature values at the experimental field ranged from 18.3°C to 40.2°C. The average external air relative humidity was 64.2% and the average wind velocity was 2.0 ms^{-1}. The monthly value of cumulative solar radiation was 643.4 MJ m^{-2} and 343.9 MJ m^{-2} on a horizontal and on a vertical plane, respectively.

During August 2016 the field was characterized by values of the external air temperature ranging from 16.2°C to 39.2°C. The average external air relative humidity was 62.5% and the average wind velocity was 2.2 ms^{-1}. The monthly value of cumulative solar radiation was 663.4 MJ m^{-2} and 333.6 MJ m^{-2} on a horizontal and on a vertical plane, respectively.

Time slots were defined, in accordance of the characteristics of the study climatic area:

- $I_{hor\ 200}$: $I_{hor} < 200$ W m^{-2}, assessed during the time period 00:00–06:30 or 17:45–23:45;
- $I_{hor\ 200\text{-}400}$: $200 \leq I_{hor} < 400$ W m^{-2}, assessed during the time period 06:45–07:30 or 16:45–17:30;
- $I_{hor\ 400\text{-}600}$: $400 \leq I_{hor} < 600$ W m^{-2}, assessed during the time period 07:45–08:45 or 15:45–16:30;
- $I_{hor\ 600\text{-}800}$: $600 \leq I_{hor} < 800$ W m^{-2}, assessed during the time period 09:00–10:15 or 14:00–15:30;
- $I_{hor\ 800}$: $I_{hor} \geq 800$ W m^{-2}, assessed during the time period 10:30–13:45.

A statistical model was developed for each time range and for each of the two green façades. The regression models were developed by using the data recorded during August 2015 for *Rhyncospermum jasminoides* and for *Pandorea jasminoides variegated*. The subscript t in eqn (1) is $t = 1, \ldots, 2976$. Tables 1–2 show the regression coefficients. Quality analysis parameters were evaluated for each model (Tables 1–2). These parameters are the coefficient of determination R^2, the adjusted coefficient of determination R_{adj}^2 and the Root-mean-square error (RMSE). R^2 and R_{adj}^2 indicate a goodness of the fit to the training data for each model. The RMSE values show that the dispersion of the data around the regression line is low.

The regression coefficients cannot be compared to each other, because they are related to different unit of measurement. A positive regression coefficient means that an increase in the value of the related predictor will increase the temperature difference.

The regression coefficients $\hat{\beta}_2$ and $\hat{\beta}_3$, which are related to the external air temperature predictor, are positive and negative. Thus, the contribution of the predictors $X_{1,\ t}$ and $X_{1,\ t-1}$ partially compensate each other. Their contribution is not significant for $I_{hor\ 400\text{-}600}$.

The coefficients $\hat{\beta}_{12}$ and $\hat{\beta}_{13}$ are characterised by absolute values higher than $\hat{\beta}_4$ and $\hat{\beta}_5$, when all exist. This shows a greater dependence of the difference of the external surface temperature between the control wall and the green façades on the vertical than horizontal solar radiation.

Table 1: Estimated regression coefficients and quality model parameters for the green façade covered with *Rhyncospermum jasminoides*, for August 2015.

Predictors variables		Y_{t-1}	$X_{1,t}$	$X_{1,t-1}$	$X_{2,t}$	$X_{2,t-1}$	$X_{3,t}$	$X_{3,t-1}$	$X_{4,t}$	$X_{4,t-1}$	$X_{5,t}$	$X_{5,t-1}$	$X_{6,t}$	$X_{6,t-1}$			
			External Air Temperature	External Air Temperature	External solar radiation horizontal	External solar radiation horizontal	External Air relative Humidity	External Air relative Humidity	Wind Velocity	Wind Velocity	Wind Direction	Wind Direction	External solar radiation vertical	External solar radiation vertical			
	Estimated regression coefficient														**Quality model parameters**		
	$\hat{\beta}_0$	$\hat{\beta}_1$	$\hat{\beta}_2$	$\hat{\beta}_3$	$\hat{\beta}_4$	$\hat{\beta}_5$	$\hat{\beta}_6$	$\hat{\beta}_7$	$\hat{\beta}_8$	$\hat{\beta}_9$	$\hat{\beta}_{10}$	$\hat{\beta}_{11}$	$\hat{\beta}_{12}$	$\hat{\beta}_{13}$	**R^2**	**R_{adj}^2**	**RMSE**
Time slot I_{hor} 200	−0.1941	0.9285	0.1493	−0.1476			0.0041	−0.0032					0.0015	−0.0024	0.99	0.99	0.06
I_{hor} 200–400	−0.2305	0.8934	0.1092	−0.1083	0.0005		0.0005								0.99	0.99	0.11
I_{hor} 400–600	−0.3051	0.9000			0.0009		0.0010		−0.0429	0.0399					1.00	1.00	0.13
I_{hor} 600–800	0.0000	0.9155	0.1155	−0.1207	0.0003				−0.0253				0.0012		1.00	1.00	0.12
I_{hor} 800	0.0000	0.9369	0.2277	−0.2293					−0.0303		−0.0005		0.0019	−0.0004	1.00	1.00	0.14

Table 2: Estimated regression coefficients and quality model parameters for the green façade covered with *Pandorea jasminoides variegated*, for August 2015.

Predictors variables		Y_{t-1}	$X_{1,t}$	$X_{1,t-1}$	$X_{2,t}$	$X_{2,t-1}$	$X_{3,t}$	$X_{3,t-1}$	$X_{4,t}$	$X_{4,t-1}$	$X_{5,t}$	$X_{5,t-1}$	$X_{6,t}$	$X_{6,t-1}$			
			External Air Temperature	External Air Temperature	External solar radiation horizontal	External solar radiation horizontal	External Air relative Humidity	External Air relative Humidity	Wind Velocity	Wind Velocity	Wind Direction	Wind Direction	External solar radiation vertical	External solar radiation vertical			
	Estimated regression coefficient														**Quality model parameters**		
	$\hat{\beta}_0$	$\hat{\beta}_1$	$\hat{\beta}_2$	$\hat{\beta}_3$	$\hat{\beta}_4$	$\hat{\beta}_5$	$\hat{\beta}_6$	$\hat{\beta}_7$	$\hat{\beta}_8$	$\hat{\beta}_9$	$\hat{\beta}_{10}$	$\hat{\beta}_{11}$	$\hat{\beta}_{12}$	$\hat{\beta}_{13}$	R^2	R_{adj}^2	RMSE
Time slot I_{hor} 200	–0.4529	0.8696	0.1336	–0.1243	0.0014	–0.0011	0.0011								0.91	0.91	0.18
I_{hor} 200–400	0.0000	0.8625	0.1211	–0.1251	0.0008			0.0010	0.0558	–0.0603	–0.0004				0.98	0.97	0.26
I_{hor} 400–600	–0.2391	0.8577			0.0010		–0.0013	0.0020							0.98	0.98	0.24
I_{hor} 600–800	–0.3942	0.8819	0.1353	–0.1250					–0.0529				0.0022		0.99	0.99	0.18
I_{hor} 800	0.0000	0.9056	0.2465	–0.2458					–0.0622		–0.0008		0.0020		1.00	1.00	0.19

In relation to the external air relative humidity, the regression coefficients $\hat{\beta}_6$ and $\hat{\beta}_7$ are not significant for $I_{hor\ 600\text{-}800}$ and for $I_{hor\ 800}$. The dependence on the wind variable is expressed by the $\hat{\beta}_8$ and $\hat{\beta}_9$ (wind velocity) and $\hat{\beta}_{10}$ and $\hat{\beta}_{11}$ (wind direction). No dependence on the wind characteristics results for $I_{hor\ 200}$ for both green façades.

Each model, which was developed using data concerning August 2015 data, was then applied to predict the performance of the green façades in August 2016. Figures 2 and 3 show, as a result, the difference between the temperature of the external surface on the control wall and on the green walls observed during the experimental test and the difference of temperature values predicted with the regression models, during 25–31 August 2016.

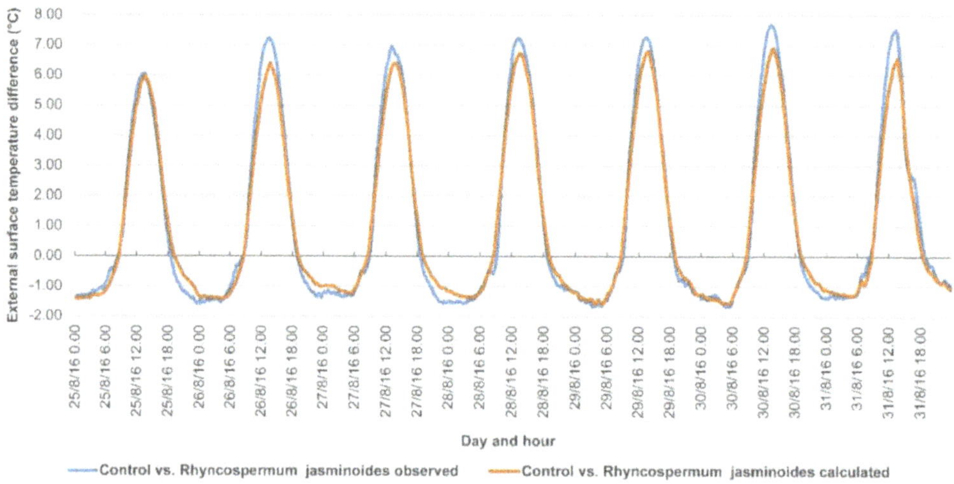

Figure 2: Difference of the external surface temperature between the control wall and the green façade covered with *Rhyncospermum jasminoides*: data observed at the experimental field and values calculated by using the regression models; 25-31/08/2016.

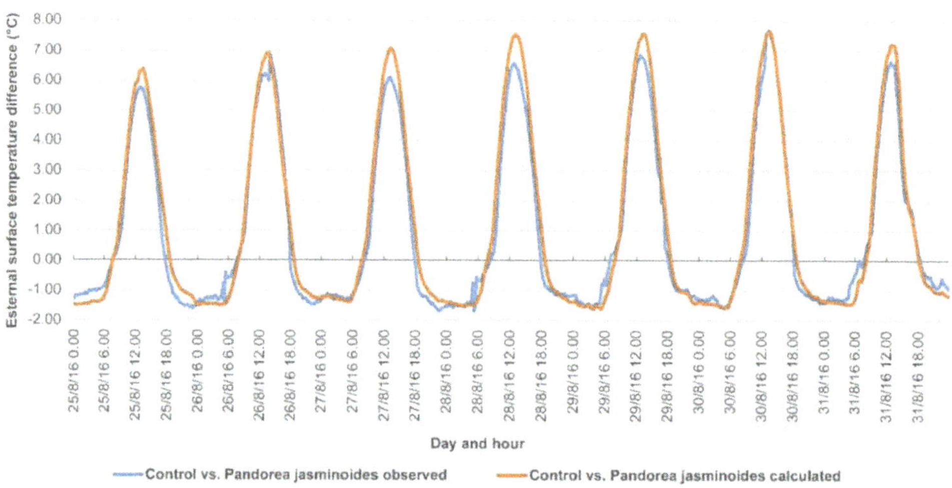

Figure 3: Difference of the external surface temperature between the control wall and the green façade covered with *Pandorea jasminoides variegated*: data observed at the experimental field and values calculated by using the regression models; 25-31/08/2016.

The models show a good predicting performance. A one-way ANOVA analysis at a 95% probability level was developed for the selected days of the testing month. It evaluated the mean values of the experimental data and the calculated data; no statistically significant differences were found for both the green façades.

The minimum difference between the external surface temperature of the control wall and the green façade covered with *Rhyncospermum jasminoides* (Fig. 2) was -1.66°C and -1.58°C for the experimental values and the predicted values, respectively. The maximum difference was 7.69°C and 6.91°C for the experimental values and the predicted values, respectively.

The minimum difference between the external surface temperature of the control wall and the green façade covered with *Pandorea jasminoides variegated* (Fig. 3) was −1.68°C and −1.62°C for the experimental values and the predicted values, respectively. The maximum difference was 7.67°C and 7.64°C for the experimental values and the predicted values, respectively.

4 CONCLUSIONS

Vertical greenery popularity is growing due to their high potential in being used as a sustainable solution for improving the thermal performance of building envelopes in the sector of residential and rural constructions. In summer in Mediterranean climate conditions, the thermal performance of the green façades still needs further research study and in particular experimental data on long periods. The assessment of the plants suitable for the Mediterranean context and of their performance is necessary for choosing the most efficient solution.

This study explores the thermal performance of two green façades that were built in Bari, South Italy. *Rhyncospermum jasminoides* and *Pandorea jasminoides variegated* were selected for covering the experimental walls. The application of the green cover allowed reducing the daylight temperatures on the external surface of the shielded walls during warm days. The experimental data were used for developing a multiple regression equation regarding the dependence of the difference of external surface temperature between the green façades and the control wall and the weather conditions. The model was trained on August 2015 data. The predictive model can be useful for predicting the external surface temperature reduction by the green cover. Its predicting ability was tested on August 2016. The model shows a good predicting performance, resulting a useful tool in comparing the efficacy of the two vertical greenery solutions in similar Mediterranean climatic contexts.

ACKNOWLEDGEMENTS

The contribution to programming and conducting this research must be equally shared between the Authors.

The present work has been carried out under the ‘Piano triennale della Ricerca 2015–2017 nell’ambito del Sistema Elettrico Nazionale, Progetto D.1 ‘Tecnologie per costruire gli edifici del futuro’, Research activity: “Analisi di tecniche di raffrescamento sostenibili applicabili in edifici civili e in edifici serra” Piano Annuale di Realizzazione (PAR) 2017’, Accordo di Programma Ministero dello Sviluppo Economico – ENEA funded by the Italian Ministry of Economic Development.

REFERENCES

[1] Norton, B.A., Coutts, A.M., Livesley, S.J., Harris, R.J., Hunter, A.M. & Williams, N.S.G., Planning for cooler cities: A framework to prioritise green infrastructure to mitigate high temperatures in urban landscapes. *Landscape and Urban Planning*, **134**, pp. 127–138, 2015. https://doi.org/10.1016/j.landurbplan.2014.10.018

[2] Pérez, G., Coma, J., Martorell, I. & Cabeza, L.F., Vertical Greenery Systems (VGS) for energy saving in buildings: A review. *Renewable and Sustainable Energy Reviews*, **39**, pp. 139–165, 2014. https://doi.org/10.1016/j.rser.2014.07.055

[3] Tan, C.L., Wong N.H. & Jusuf, S.K., Effects of vertical greenery on mean radiant temperature in the tropical urban environment. *Landscape and Urban Planning*, **127**, pp. 52–64, 2014. http://dx.doi.org/10.1016/j.landurbplan.2014.04.005

[4] Vox, G., Blanco, I., Campiotti, C.A., Giagnacovo G. & Schettini, E., Vertical green systems for buildings climate control. *Proceedings of the 43rd International Symposium – Actual Tasks on Agricultural Engineering*, Sveučilište u Zagrebu, Agronomski fakultet, Zavod za mehanizaciju poljoprivrede. Opatija, Croatia, 24-27/02/2015, pp. 723–732, 2015.

[5] Vox, G., Blanco, I., Fuina, S., Campiotti, C.A., Scarascia Mugnozza, G. & Schettini, E., Evaluation of wall surface temperatures in green facades. *Proceedings of the Institution of Civil Engineers – Engineering Sustainability,* **170(6)**, pp. 334–344, 2017. https://doi.org/10.1680/jensu.16.00019

[6] Berardi, U., Ghaffarianhoseini, A.H. & Ghaffarian Hoseini, A., State-of-the-art analysis of the environmental benefits of green roofs. *Applied Energy*, **115**, pp. 411–428, 2014. http://dx.doi. org/10.1016/j.apenergy.2013.10.047

[7] Fernandez-Cañero, R., Emilsson, T., Fernandez-Barba, C. & Herrera Machuca, M.A., Green roof systems: a study of public attitudes and preferences in southern Spain. *Journal of Environmental Management*, **128**, pp. 106–115, 2013. http://dx.doi.org/10.1016/j.jenvman.2013.04.052

[8] Santamouris, M., Cooling the cities – a review of reflective and green roof mitigation technologies to fight heat island and improve comfort in urban environments. *Solar Energy*, **103**, pp. 682–703, 2012. http://dx.doi.org/10.1016/j.solener.2012.07.003

[9] Cheng, C.Y., Cheung, K.K.S. & Chu, L.M., Thermal performance of a vegetated cladding system on facade walls. *Building and Environment*, **45(8)**, pp. 1779–1787, 2010. https://doi.org/10.1016/j.buildenv.2010.02.005

[10] Jim, C.Y. & Tsang, S.W., Biophysical properties and thermal performance of an intensive green roof. *Building and Environment*, **46(6)**, pp. 1263–1274, 2011. https://doi.org/10.1016/j.buildenv.2010.12.013

[11] Köhler, M. & Poll, P.H., Long-term performance of selected old Berlin greenroofs in comparison to younger extensive greenroofs in Berlin. *Ecological Engineering*, **36(5)**, pp. 722–729, 2010. https://doi.org/10.1016/j.ecoleng.2009.12.019

[12] Perini, K., Ottelé, M., Fraaij, A.L.A., Haas E.M. & Raiteri, R., Vertical greening systems and the effect on air flow and temperature on the building envelope. *Building and Environment*, **46(11)**, pp. 2287–2294, 2011. https://doi.org/10.1016/j.buildenv.2011.05.009. https://doi.org/10.1016/j.buildenv.2011.05.009

[13] Safikhani, T., Abdullah, A.M., Ossen, D.R. & Baharvand, M., A review of energy characteristic of vertical greenery systems. *Renewable and Sustainable Energy Reviews*, **40**, pp. 450–462, 2014. https://doi.org/10.1016/j.rser.2014.07.166

[14] Gago, E.J., Roldan, J., Pacheco-Torres R. & Ordóñez, J., The city and urban heat islands: A review of strategies to mitigate adverse effects. *Renewable and Sustainable Energy Reviews*, **25**, pp. 749–758, 2013. https://doi.org/10.1016/j.rser.2013.05.057

[15] Cameron, R.W.F., Taylor J.E. & Emmett, M.R., What's 'cool' in the world of green façades? How plant choice influences the cooling properties of green walls. *Building and Environment*, **73**, pp. 198–207, 2014. https://doi.org/10.1016/j.buildenv.2013.12.005

[16] Yang, F., Yuan, F., Qian, F., Zhuang, Z. & Yao J., Summertime thermal and energy performance of a double-skin green facade: A case study in Shanghai. *Sustainable Cities and Society*, **39**, pp. 43–51, 2018. https://doi.org/10.1016/j.scs.2018.01.049

[17] Raji, B., Tenpierik, M.J. & van den Dobbelsteen, A., The impact of greening systems on building energy performance: A literature review. *Renewable and Sustainable Energy Reviews*, **45**, pp. 610–623, 2015. https://doi.org/10.1016/j.rser.2015.02.011

[18] Medl, A., Stangl, R. & Florineth, F., Vertical greening systems – A review on recent technologies and research advancement. *Building and Environment*, **125**, pp 227–239, 2017. ISSN 0360-1323. https://doi.org/10.1016/j.buildenv.2017.08.054

[19] Pérez-Urrestarazu, L., Fernández-Cañero, R., Franco-Salas A. & Egea, G., Vertical Greening Systems and Sustainable Cities. *Journal of Urban Technology*, **22(4)**, pp. 65–85, 2016. https://doi.org/10.1080/10630732.2015.1073900

[20] Lin, H., Xiao, Y. & Musso, F., Shading Effect and Heat Reflection Performance of Green Façade in Hot Humid Climate Area: Measurements of a Residential Project in Guangzhou, China. *IOP Conference Series: Earth and Environmental Science*, **146,** p. 012006, 2018. https://doi.org/10.1088/1755-1315/146/1/012006

[21] Kontoleon, K.J. & Eumorfopoulou, E.A., The effect of the orientation and proportion of a plant-covered wall layer on the thermal performance of a building zone. *Building and Environment*, **45**, Suppl. 5, pp. 1287–1303, 2010. http://dx.doi.org/10.1016/j.buildenv.2009.11.013

[22] Giordano, R., Montacchini, E., Tedesco S. & Perone, A., Living Wall Systems: A Technical Standard Proposal. *Energy Procedia*, **111**, pp. 298–307, 2017. http://dx.doi.org/10.1016/j.egypro.2017.03.093

[23] Eumorfopoulou, E.A., & Kontoleon, K.J., Experimental approach to the contribution of plant-covered walls to the thermal behaviour of building envelopes. *Building and Environment*, **44(5)**, pp. 1024–1038, 2009. https://doi.org/10.1016/j.buildenv.2008.07.004

[24] Campiotti, C.A., Schettini, E., Alonzo, G., Viola, C., Bibbiani, C., Scarascia Mugnozza, G., Blanco, I. & Vox, G., Building green covering for a sustainable use of energy. *Journal of Agricultural Engineering,* **44(2s)**, pp. 253–256, 2013. https://doi.org/10.4081/jae.2013.292

[25] Blanco, I., Schettini, E., Scarascia Mugnozza, G., Campiotti, C.A., Giagnacovo, G. & Vox, G., Vegetation as a passive system for enhancing building climate control. *Acta Horticulturae,* **1170**, pp. 555–562, 2017. https://doi.org/10.17660/actahortic.2017.1170.69

[26] Vox, G., Blanco, I. & Schettini, E., Green façades to control wall surface temperature in buildings. *Building and Environment*, **129**, pp. 154–166, 2018. https://doi.org/10.1016/j.buildenv.2017.12.002

[27] Hunter, A.M., Williams, N.S.G., Rayner, J.P., Aye, L., Hes, D. & Livesley, S.J., Quantifying the thermal performance of green façades: a critical review. *Ecological Engineering,* **63**, pp. 102–113, 2014. http://dx.doi.org/10.1016/j.ecoleng.2013.12.021

[28] UNI EN 1745 2012. Masonry and masonry products - Methods for determining thermal properties. UNI – Ente Nazionale Italiano di Unificazione – Italian Organization for Standardization; IT.

[29] Fumo, N. & Rafe Biswas, M.A., Regression analysis for prediction of residential energy consumption. *Renewable and Sustainable Energy Reviews*, **47**, pp. 332–343, 2015. DOI: 10.1016/j.rser.2015.03.035 https://doi.org/10.1016/j.rser.2015.03.035

[30] Yildiz, B., Bilbao J.I. & Sproul, A.B., A review and analysis of regression and machine learning models on commercial building electricity load forecasting. *Renewable and Sustainable Energy Reviews*, **73**, pp. 1104–1122, 2017. https://doi.org/10.1016/j.rser.2017.02.023

COMPARING URBAN FOOD SYSTEMS BETWEEN TEMPERATE REGIONS AND TROPICAL REGIONS-INTRODUCING URBAN AGROFORESTRY IN TEMPERATE CLIMATES THROUGH THE CASE OF BUDAPEST

P.G. DE LINARES
Department of Landscape Architecture and Regional Planning, Szent István University, Budapest.

ABSTRACT

The aim of this paper is firstly to assess what makes urban agriculture more successfully integrated in some cities than others. Secondly, to introduce agro-ecology practices in public green open spaces and community gardens through a landscape assessment and a map-based comparison analysis. The main problem is to motivate the planners to integrate urban agriculture in the Urban plan. Therefore, these green structures with their benefits cannot be part of the city's landscape and dynamic. The sustainability of food systems depends on the planning strategy of the city and the governance policies. Whilst urban agroforestry is well applied in Tropical Climate, it has not been fully explored in Temperate climate. This practice could have multiple functions in the Temperate Region and become a sustainable land use thanks to agro-ecology principles. After defining urban agroforestry for Temperate Regions, a methodology to find the best spaces to introduce agro-ecology practices will be evaluated through the case of Budapest where a green infrastructure plan has recently been launched and an agroforestry project is being initiated. This paper concludes that urban agroforestry is a sustainable land use that can better integrate food systems in the city.

Keywords: agro-ecology, food security, landscape architecture, sustainable food system.

1 INTRODUCTION

As urban agriculture is expanding around the world (FAO [1]), it occurs that implementing this practice in landscape architecture and town planning is necessary. Cities are rapidly growing and are expected to spread more intensively in the future (EEA [2]). This threatens the green spaces, agricultural lands, natural habitats and ecosystems. Loss in biodiversity is caused by fragmentation of natural habitats and anthropic activities such as intensive farming and monoculture. Planners have denied their role in planning food systems and providing food in cities, creating an imbalance in food provisioning and in the landscape, 'Urban planners might justify this 'puzzling omission' by claiming that the food system is largely a rural issue and therefore beyond the scope of the urban planning agenda.' (Morgan [3]). With upcoming climate change scenarios human beings are more vulnerable, socially, economically and environmentally. It is time governments and planners consider urban agriculture as a solution for building resilient cities and landscapes. Uban agriculture can be defined as an intentional and deliberate practice for reconstruction of ecosystems with strong social participation for enhancing food security in cities. Based on a scientific research and a literature review, different terms could be found to define food systems: *edible landscapes*, *consumption patterns, edible green infrastructure, continuous productive urban landscapes, new urban agriculture.* Urban agriculture can be classified into a typology of structures: *allotment gardens, urban farms, green roofs, green walls, urban orchards, vertical farms.* However, policies around urban agriculture are different across the world. The question is how to integrate these systems in the urban landscape as a full land use. According to Wiskerke, there are three ways to succeed in the planning of food systems: the creation of a food policy

DOI: 10.2495/DNE-V13-N4-395-406

council, a municipal department of food and placing food as the responsibility of the planning department (Wiskerke [4]). Edible landscapes enhance food security in the city, reduce dependence on importations and provide social and economic benefits. The environmental benefits of urban agriculture are still under debate as there is a disparity in the world's access to good quality soils, water and air.

In this study, we define sustainable food systems as a way of growing edibles with agro-ecology principles and on a secured long-term land to enhance food security, for a consumption of high-quality food in equal access. The economic outcomes are also to be considered in sustainable food systems, but this is not covered in the paper. The aim of this research is to assess how agroforestry could be integrated in Temperate Urban Climates and the purposes this practice would have. Agroforestry is defined as 'a collective name for land-use systems and technologies where woody perennials are deliberately used on the same land-management units as agricultural crops and/or animals, in some form of spatial arrangement or temporal sequence. In agroforestry systems there are both ecological and economical interactions between the different components' (Ramachandra [5]). There must be a significant interaction between the woody and nonwoody components of the system, either ecological and/or economical. The questions are why should we turn to urban agroforestry practice in urban environments and how to plan this food system ? The aim of this paper is firstly to assess what makes urban agriculture more successfully integrated in some cities than others. Secondly, to introduce agro-ecology practices in public green open spaces and community gardens through a landscape assessment and a map based comparison analysis. Case studies, articles and observations were used to understand the policies behind urban agriculture in the northern and southern hemisphere. A literature review highlights the benefits of agroforestry systems and urban agroforestry practices. After defining urban agroforestry for Temperate Regions, the analysis focuses on the case of Budapest where a landscape assessment helps to understand the purposes of agroforestry in the city and ways in planning. This paper concludes that urban agroforestry is a sustainable land use that can better integrate food systems in the city.

2 SUCCESS AND DISPARITIES IN URBAN AGRICULTURE

After the UN Summit of Rio de Janeiro on Sustainable Development, the City of Montreal established a local planning program called *eco-neighbourhoods* in 1995, in the frames of a Local Agenda (Reyburn [6]). This has empowered the districts and citizens to initiate and plan green actions in their neighbourhoods. The two main initiatives driven by the citizens are intensive urban horticulture and green laneways. Since 2002, the eco-neighbourhoods are financed by their districts (REQ [7]). Whilst these were mainly created to raise social awareness on the environment it occurs that the main objectives driven by the citizens have become social. Initially there were four policy measures: 1. education and action program in environment 2. Community based actions 3. Encourage citizens to create environmental actions themselves and get involved. There were also four programs: 1. Clean neighbourhoods 2. Waste management 3. Beauty and aesthetics 4. Nature in the city. Today, there are 16 econeighbourhood programs in Montreal. Urban horticulture was motivated for insuring food security and more social equality between the districts as in 2006, 36% of children lived under the poverty line (Boudreau [8]). The responsibles of the program are employed by the district and are offered a place to gather and run their projects with volunteers. The intensive gardening spread to the housing plots of duplex, triplex and quadruplex separated by laneways and small backgardens, typical of Montreal's urban fabric. Urban gardening is part of the urban fabric and dynamic. The main successes of this program are the creation of jobs, internships and a strong knowledge exchange through educational programs. Community gardens, schools and even universities organise workshops, activities and international programs to teach horticulture and food systems planning in the cities and their peripheries. From participating in the School of Urban Agriculture Week in Montreal's university UQAM in 2013, it appears that the planning concept of Small is Beautiful is the motor of these initiatives. Urban agriculture in semi-public plots is animated by horticultural and gardening instructors employed by the district. These are responsible for the management of the plots and volunteers. This helps in building collective activities and fare sharing of harvests. In 2013, it was reported that 42% of the inhabitants in Montreal practiced urban agriculture, which means 800 000 people. It is also important to mention that this gardening is not just expanding in poor quarters but also in up-market neighbourhoods in the core of the City through a green network between the laneways. These green laneways and gardens have become urban green trails, which can be followed on maps. In time, the metropolitan map of the green and blue network shows that there will be a connection between the locally planned green network and the metropolitan plan. Also, greening the laneways was a way to revive these paths and change the bad reputation and image they had (Beaudet [9]). Along these can be found gardens, art and recreational activities. Green Laneways help in reducing departures to the suburbs as they provide the same benefits: a green space in front of homes with safety for children to play in (Guelper [10]). It is not just a plan for green spaces in the city but also a social program. Thanks to a public consultation about the importance of urban agriculture in Montreal in 2012, this practice is part of the Urban Plan, the Sustainable Development Plan and carried out through other major programs of the City. The fact that these programs were carried out by motivated citizens and organised with animators are important elements for the success of public and semi-public edible gardens.

In Nice, South of France, a community garden was founded by an NGO 'Graines de Fermiers'. From experience and observations, it seemed that the lack of animator lead to long debates between volunteers and a more individualist vision with low sharing and communication. There was a problem with the organisation in management of green wastes, watering and even share of the plots. The garden is isolated from the rest of the city and is difficult to access to or even notice. Community gardeners in France are often created and managed by NGOs who depend on governmental funding to maintain their activities with volunteers.

The case of Havana stresses the importance of education and governmental implication for sustainable food systems. In Havana a Policy for Urban Agriculture has been running since 1997 and a decentralised law for urban agriculture was established (Observatoire Villes Inclusives [11]). The government has built strong educational programs to encourage and teach the urban gardeners to work with active organic methods. The importance of Urban Agriculture in Havana is proven by the creation of a Department of Urban Agriculture. Thanks to a governmental change on land use, and the autoconsumo plan initiated in the 1980s (Gonzalez [12]) there is a high accessibility to lands for gardening. This has lead to a rapid expansion of this practice in the city and self- food provisioning for families and employment on farms and less resource demanding needs and transport.

In Dar Es Salaam, urban agriculture is used as a measure to end settlements growth in flood risk areas (Howorth [13]). This has lead to a better management of the landscape and brought food closer to families and employment.

In Budapest, urban agriculture was very strong during the two World Wars. Until the 1980s, Bulgarian farmers on the Csepel Islands, District (XXI) provided food to the city. They helped in reducing hunger as they sold their organic products from the island and on markets across the city (Körösi [14]). Only a small number of farmers remain today. Since the integration in the EU, importation has slowly ended this practice on the island. Hus a new housing estate was built in this island destroying around 5 hectares of sandy grassland vegetation (Tenk [15]). The agglomeration is expanding with rapid urban sprawl since late 1980s, causing the loss of agricultural lands and natural habitats. But small investment is made on green spaces and urban agriculture in Budapest. A new project initiated by the Government seeks to develop cultural infrastructures on the City Park Városliget. This despite the fact that Budapest only offers 6 m^2 recreational green space per capita (Green Infrastructure Concept, BFVT Ltd [16]). Some community gardens have been initiated and expanding since 2010. The problem is that community gardeners in Budapest, like other cities, depend on short-term contracts with tertiary sector companies. There are 37 edible gardens in Budapest (Faczanyi [17]). These are school gardens, allotment gardens or community gardens. They are founded either by civil initiative or by the contemporary architecture center (KEK) and the NGO VKE. The first garden was created in 2010 on a 400 m^2 land but closed in 2013. An interview with the manager of Grundkert revealed that they had to move from their land 3 times in the last 5 years due to ending of contract with partners. Under these circumstances it is impossible to grow sustainable food in the city and to revive ecosystems in the landscape because there is not enough time. The community gardens are fragmented from the other functions in the city. Short-term contracts with companies cannot provide good conditions for urban agriculture. Between 2010 and 2017, six community gardens were closed.

What makes urban agriculture more successfully implemented in some cities than others is firstly the involvement of the government and the educational support behind this practice and secondly, the sustainability of the ecosystem in the plot. Scenarios for climate change show a need to build resilient agricultural systems, both in cities and rural areas. Agroforestry is a traditional agricultural practice, which has been forgotten in the Temperate Regions due to modern technologies and industrialisation. However, the practice has strongly remained in Tropical and Subtropical Climates. Today it is being revisited by agronomists showing that the value of biodiversity in agriculture has an impact on food productivity and economy. Whilst urban agroforestry is well applied in Tropical Climate, it has not been fully explored in Temperate Climate. This practice could have multiple functions in the Temperate Region and become a sustainable land use thanks to agro-ecology principles.

3 URBAN AGROFORESTRY

Different agroforestry systems exist: industrial plantations, community woodlots, farm woodlots, trees in crop land, alley farming, linear planting, shelterbelts, sequential cropping, wood pasture, protection forestry, land rehabilitation, reclamation (Burky [18]). The important characteristic in all these systems is the use of *multipurpose trees and shrubs,* meaning a choice in planting for several purposes, products, benefits and services. Experiments have been collected about the benefits of trees on crop yields. Chinese farmers plant Paulownia species as their shape provides enough light and shade for undergrowing tea plants (Chinese Academy of Forestry [19]). Plants can bring mutual benefits to each other. The research of these agronomic interactions between plants and animals is called agro-ecology (Gliessman [20]). More diversity means more ecosystem services, which means more sustainable landscapes. Trees bring many benefits to the plot. They lower temperatures on the plot, reducing need in water. They provide organic matter, reducing need in fertilizers. They attract pollinators and provide shelter to fauna. Old trees are an important element for biodiversity as they attract birds and other animals and insects, which are useful to repel parasites and balance the ecosystem. The most valuable service provided by trees is the protection and revitalisation of soils. The roots help in stabilizing the soil and in smoothing the soil. The wood and leaves help in enriching the soil and creating biologic activity (Bourguignon [21]). Some trees have been more beneficial than others, for instance Poplars, which are intensively used in the south of France amongst vineyards (AFAF [22]). The ecosystem of the soil is a most valuable resource. Its organisms regulate the circulation of air and water, the temperature and the access to nutrients. Must be mixed families, species and stratas of plants to gain multiple benefits from soils and reduce vulnerability to climate change and disease. The national research in science INRS and the French organisation for agroforestry AFAF finance many farmers to experiment agroforestry. A wider project is carried out through a cooperative between municipalities: the Biovallée, 'Organic Valley'. It has grown a multifunctional agro and socio-ecological landscape with new employment, educational programs and markets. But despite the growing use of agroforestry in rural landscapes, this practice is still underdeveloped in urban landscapes.

Through case studies, it was found that the main varieties of edibles were domestic crops such as: radish, carrots, tomatoes, beans, cucumbers, lettuce and peppers. Rarely shrubs, trees, berries or even perennials. Some initiatives show an interest in valuing fruit trees but these are carried out separately. For example the world map 'Falling Fruit' is a US cooperation inviting volunteers to map the public fruit trees in their cities. In Montreal, there is also a city map on which garden owners invite people to help harvest their fruit trees to avoid waste. But they have a separate program for planting public trees with citizens. In the south of France more research is conducted for diversifying species in renewed woodlands or reforestation programs in the frames of agroforestry by mixing wide canopy trees with fruit trees, 'Boisement mélangé' (Becquey [23]). By reviewing green strategy plans and food plans in northern countries, it appears that there is often a separation between urban forests, urban farms and urban agriculture. And a choice between growing fruit trees, ornamental trees and annual crops. In Budapest, the green infrastructure plan doesn't mention any green space for food provisioning and community gardens. A wider place could be given to trees in edible gardens and more connections could be made between food systems, green spaces and other functions in the city. In Havana, the Policy for Urban Agriculture includes growing of trees (Observatoire Villes Inclusives [11]). They have a forestry program, *Mi Programma Verde*. But reports stated that there was a lack of diversity in seeds and species (Gonzalez [12]). Therefore, there is a need in designing food systems with more diversity of species and stratas.

A more extensive agroforestry system is growing interest in gardeners: *Forest-gardens*. Traditional forest-gardens have been growing in home gardens since ancestral times in Tropical and Subtropical countries. They all have different names and are attached to different values and religions, for example in Java, Nepal and Sri Lanka. Urban agroforestry is a well- established practice in the Pacific Islands. Intensive agroforestry systems are grown on idle lands in these by families to self-provide food for their homes. These are expanding in cities in response to increased urban density and low proximity with rural areas (UNU [24]). On these lands people grow a high diversity of edibles, medicinal plants, wood and biomass. For example Palm trees are inter-cropped with coffee plants. The Kingdom of Tonga has registered this practice in the strategic plan for resilience and adaptation to climate change (UNDP [25]). The plan includes a section for agro-ecology practice for sustainable agriculture and food security: 1. Promote conservative cultivation such as minimum tillage, green tillage, vegetative mulching, etc. 2. Promote the use of bush fallow, planted legume fallow, etc. 3. Promote conservative input of the correct mineral fertilizer at the right amount in combination with appropriate organic fertilizer 4. Promote conservation cultivations of contour boundary hedgerows, terracing, mulching, green tillages, planted fallow, etc. 5. Promote alternative local species/ varieties or breeds 6. Promote introduction of exotic varieties or breeds suitable for the more humid warmer climates 7. Promote integrated pest management strategies with resistant varieties, biological (UNDP [25]). Other worldwide projects are carried out with forest gardening practices. For example, *Greening the Desert* is an experiment for growing food and providing water in arid regions. This shows that agroforestry is not only a sustainable way to grow food but also a way to preserve landscapes and reduce environmental and social vulnerability. It also increases the production on a small space.

Forest gardens were introduced in Temperate Climates by Robert Hart in 1960 who modelized the concept in England. Today the leading reference is Martin Crawford who has 500 varieties of perennial edibles in 1 ha woodland in England [26]. He uses 7 to 9 plant stratas, which all have a purpose in the garden. Ground covers are plants, which expand horizontally and help in maintaining humidity in the soil, as well as smoothing the soil so other plants can expand their roots to access to water and nutrients. They can also attract pollinators. Mushrooms help in creating an active organic soil. Vines or climbing plants can be grown on trees. Herbs help in attracting pollinators, can help in fertilizing the soil and repelling parasites. Shrubs attract birds that can also help in repelling parasites. Fruit trees can be supports for climbing plants. Wide canopy trees create shelter for pollinators and regulate the temperature. Some people even create ponds and have aquatic plants in their gardens to create a fresh micro-climate. This model implies a full design strategy and preparation of land for companion planting and to create a self-sufficient ecosystem. Not only should be mixed plant stratas but also plant families and species.

This practice is growing across European homegardens and private woodlands but rarely in public spaces. Swedish researchers Clark and Nicholas introduced the concept of 'Urban Food Forest' (UFF) in 2013 [27]. This practice is mainly based in homegardens, community orchards and perennial urban agriculture. In the United States can also be found Woody Perennial Polyculture (Center of Agroforestry [27]). An agroforestry community garden was experienced with Subarctic First Nations in Ontario (Spiegelaar [28]). But still the potential of this practice has not fully been studied. For the Temperate Climates agroforestry can be defined as the introduction of multipurpose trees and other woody perennials in the urban landscape for environmental, economic and social benefits. The purposes learnt from agroforestry could have an impact on the urban environment and the landscape, like reducing heat stress and providing more diverse yields.

4 INTEGRATING AGROFORESTRY IN THE URBAN LANDSCAPE, THE CASE OF BUDAPEST

An interest in agroforestry is growing in Hungary. Experimental plots were funded by the European program AGFORWARD in the south of Hungary, in Fajsz, and managed by a cooperative [29]. There is also a research department on agroforestry in the University of West Hungary in Sopron. A few families and NGOs are growing forest-gardens in towns and settlements outside of Budapest. For instance, a forest- garden is managed by an NGO and students on the campus of the University of Gödöllő. In October 2017, for the benefit of this research, a conference and workshop were organised in partnership with the Center of Contemporary Architecture of Budapest (KEK) and the NGO Cargonomia to present the concept of urban agroforestry and find potential plots. The event gathered 50 people and a high level of motivation could be observed. The public consisted of locals, agronomists, architects, biologists, planners and representatives of district municipalities. Despite the political context, it seems like a good time to experiment agroforestry in Budapest.

Budapest is divided into two micro-climates creating two different ecosystems on both sides of the river Danube. The west side is windy and damp. The 500 meters high Buda hills are covered by protected forests of Oaks. The east side, Pest, is the beginning of the Hungarian plains the 'Alföld'. The area is dry and sandy. It used to be covered by wetlands. Few grasslands, wetlands and woodlands remain. When looking at the green intensity map of Budapest (Fig. 1), the dichotomy is clear between Buda and Pest. The West side has a much higher proportion of tree canopy and lesser density of residents than the East side. The Pest side is more developed and mineralized by grey infrastructure.

The heat map (Fig. 2) shows a direct relation between the green spaces and the temperature of the soils. Budapest has a high proportion of sealed soils of 160 m^2/capita (EEA [30]). There is low access to water for no access to the Danube. Collective housing plots have typical private inner courtyards, which are shared between neighbors. Their mineralized surface also increases heat stress in the summer. According to the analyses of the Green Infrastructure Concept of Budapest (BFVT Ltd. [16]), only 6 m^2 per capita of green open space is offered in the city, which is much lower than the World Health Organization requirement of 9 m^2 per city dweller. This shows a need in providing recreational green space in the city, especially in the East. Vegetation must be adapted to very harsh conditions, drought and coarse winters. Should also be considered that 50% of the country's homeless live in Budapest, meaning 7500 (Homeless statistics [31]). Therefore, increasing access to public food spaces seems necessary.

In Budapest, the city trees are planted and managed by Fökert a non- profit organisation. The public green spaces are managed by the districts. Since 2017, a program is funded by the government to plant 10 000 trees in Budapest. These are mainly planted in rows on central reservations of the roads and do not serve any social or environmental purposes. The problem

Figure 1: Green intensity map of Budapest.
Source: Sándor Jombach, in Green Infrastructure Concept of Budapest, BFVT Ltd., 2017.

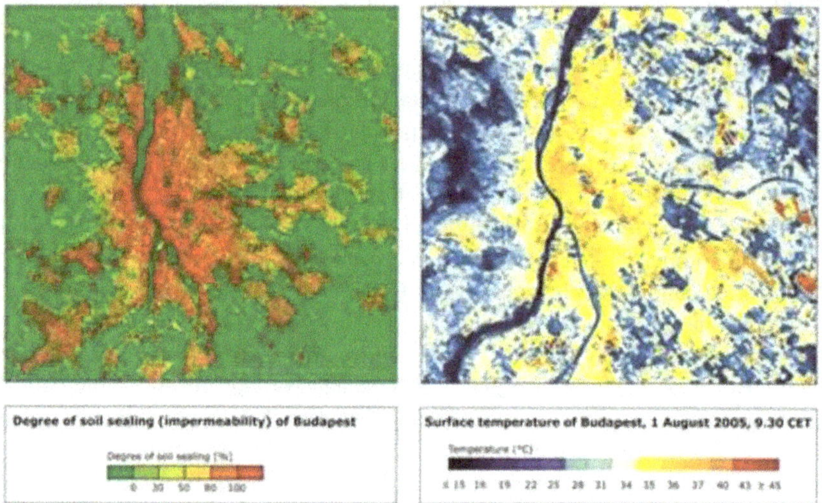

Figure 2: Map on surface temperature, Budapest.
Source: EEA, Richard Ongjerth, Péter Gábor, Sándor Jombach, 2007 and Péter Gábor, Sándor Jombach, Richard Ongjerth, 2008.

is that most young trees do not survive more than 10 years and often need to be replaced (Szaller [32]). This is due to the poor conditions of the soils, to the inaccessibility to water and nutrients and a bad preparation of the soil for the roots to develop. The main trees planted are *Acer platanoides*, *Acer negundo*, *Aesculus hippocastanum*, *Acer saccharinum*, *Acer campestre* (native), *Fraxinus excelsior*, *Ailanthus glandulosa*, *Celtis Ulmus* effuse Populus and genus of native *Sorbus* and *Tilia* (Szaller [32]). Some trees could have multipurpose in the city and be more suitable for the environmental conditions. For instance, *Sorbus* spp. has a high tolerance to urban pollution and has resilient seeds, which can grow in poor soils even after 5 years (Bouton [33]). The fruit from *Sorbus aucuparia* attract birds. Seeds spread out widely and these trees do not require much maintenance. They are also popular in home gardens for their easy level of maintenance and aesthetic value. In a wider urban planning strategy, agroforestry could be integrated in the city's landscape. Through aerial images, land use maps, geological maps, social maps, the analysis of green cover and housing plots, landscape architects can assess the best places for growing food in the city and enhancing connections between consumption patterns and other functions in the city. A map of contaminated land should be done to find safe places for growing food and assess where soils could or should be restored for this purpose. Historical maps can be used to review landscapes and ecosystems to renew in the city. The inner residential courtyards provide a space for testing agroforestry on a long-term bases because the management would be directly organised by the owners. Multipurpose trees and woody perennials could also be tested in allotment gardens.

Community gardens could be used to try agroforestry practice for environmental, social and economic outputs. By assessing these gardens with high resolution aerial images in Google Earth we could see whether trees were planted on the plot or not and if there were any other edibles than annual crop. For example, Leonardo Kert was a 1400 m^2 community garden held by the KEK. Despite the width of the area, no trees were planted. Due to an ending contract this was closed in 2017. Árnyas kert is a 1213 m^2 community garden with 29

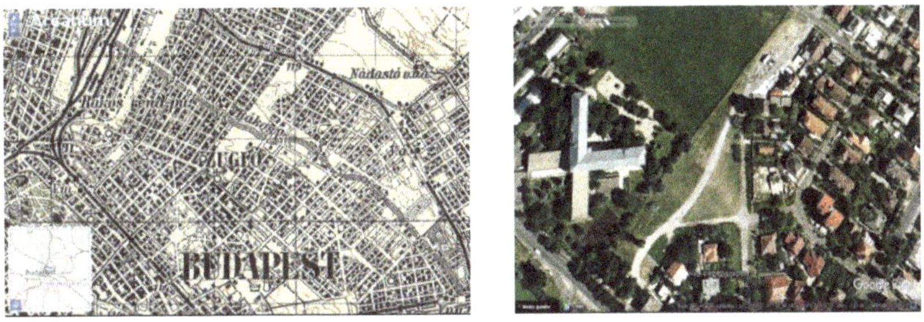

Figure 3: The evolution of a wetland along the Rakos Creek; between 1941 and 2016. *Sources*: Military Survey of 1941, Mapire.eu; Landsat 8 image, Google Earth.

gardeners. Trees were already planted on site but these are not valued and used. These gardens could be expanded and connected to wider green connections in the city. For example, public open green spaces between high residential buildings offer small-scale green connections, which could be explored for planting continuous edible landscapes.

On a larger scale, the Rakos creek is a 44-km corridor. which connects the city of Gödöllő to the Danube River in Budapest. This used to be an agricultural area and crabs used to be sold on markets directly from the stream. But due to pollution and expansion of the city the fauna was lost. The creek crosses many semi-natural spaces and dense residential neighbourhoods. It also connects different green spaces such as abandoned woodlands, agricultural lands and grasslands. Some underused sandy lands were also found (Fig. 3). An interest in reviving this landscape seems to be growing. The Green Infrastructure Concept intends to value the creek and enhance its potential as a recreational green space. An NGO 'Zöld XVII' is improving the landscape of the creek in the 17th district by planting trees with the community and raising environmental awareness campaigns. People use this area for jogging, walking their dogs and cycling. By analysing the heat map and green intensity map, it is obvious that the creek offers a great potential for recreation and reducing heat stress. It also reveals the intensity of grey infrastructure and the direct negative effect on the creek's benefits in the dense inner city.

This offers a potential green corridor where agroforestry systems could be connected to other recreational green spaces and ecosystem services. Agroforestry could be tested for its impact on renewing the landscape, the quality of the water and the comfort for people. The districts of Budapest generally offer rich mosaics. They are subdivided into diverse green patterns mixing parks, woodlands, agriculture, semi-natural areas and open green spaces between buildings. These all offer resources for combining green spaces with food systems with a close connection to people.

5 DISCUSSION

From presenting urban agriculture in universities, it seems that planners view this practice as a step backwards and a loss of potential land to develop. Thus, the lack of integration of urban agriculture in landscape design and town planning leads to fragmentation between consumption patterns, public green spaces and other urban functions. Urban gardeners often struggle to find long-term spaces to grow food. This leads to a low choice in land and sometimes poor conditions for growing sustainable food. In order to convince planners to take into account food system planning in the cities, the research of multifunctionality of food systems is necessary. The loss in biodiversity together with remaining social inequalities between cities and even between their inner districts, for access to green space, recreation or food security, shows there is a need to plan with ecosystems and create eco-social programs in the city and with peri-urban farms. As our resources in healthy food, healthy soils and healthy air have decreased and the need to adapt to climate change is a long-term matter, it is necessary to restore natural resources in cities. There is also inequality in access to fertile and uncontaminated soils, water resources and land for growing sustainable food systems. The policy behind urban agriculture has an impact on its integration in the urban landscape. In some cities, people can live from urban agriculture. In other countries like France, it is impossible because the NGOs always search for funding to support their activities and cannot create any jobs. In Budapest, community gardens are vulnerable in their partnerships with companies. To avoid fragmentation, the case of Montreal showed that smooth tools empowering the citizens to take initiatives had a positive impact on local green networks and the landscape. It has influenced people's behaviour and appropriation to space. The governmental support is essential to open lands for growing sustainable food and monitoring the conditions. With an agroecological approach better conditions could be created to grow trees and food in the city. It is important to connect urban food systems with other green spaces to increase biodiversity in the city, mix vegetation species and green patterns. Through cooperation between NGOs, green space managers and planners, food systems could be integrated to green infrastructures. The main challenge in food systems is to find good soils and to adapt to the urban context by choosing resilient edibles. More research should be done to understand the functions of species in the ecosystems to create more resistant, sustainable and diverse green structures and green spaces. It is not only about growing food but growing the right food in our cities. Educational programs are essential to gain all benefits from urban agriculture and expand the practice in a sustainable way.

6 CONCLUSION

To grow sustainable food there is a need in long-term plots with high biodiversity to create a rich ecosystem as this provides a safer environment for growing food. Soils are an important resource, not only for growing food but also for sequestering carbon. By combining trees with food systems, more benefits could be provided for the whole city with social, economic and environmental outcomes. With a landscape architecture approach, it is easier to find the best places for growing edibles in the city and enhance equal access to food and open green spaces. Many different systems exist in agroforestry but in a dense city context the most suitable system could be forest gardening as it requires less space. The forest-garden in Gödöllő could be connected to this Rakos creek green corridor through community agroforestry woodlands and agroforestry gardens. A strong cooperation should be planned between gardeners and farmers along this corridor to create a full agro-ecology program and avoid a domino effect from contamination of lands with chemicals and pesticides. The limiting factor is that urban agroforestry is a long-term process. Annuals grow faster, and therefore, it is possible to feed people quickly on a daily bases. But with more diversity in food systems across the city, it is possible to overcome this problem and adapt to the rhythm of trees.

REFERENCES

[1] Food and agriculture organization of the United Nations, FAO, 2017. http://www.fao.org/home/en/ (accessed on 3 October, 2017).

[2] European Environment Agency, EEA, 2006. https://www.eea.europa.eu/ (accessed on 5 May, 2018).

[3] Morgan, K., Feeding the city: the challenge of urban food planning. *International Planning Studies,* **14(4)**, pp. 341–348, 2009. https://doi.org/10.1080/13563471003642852
[4] Journal article: Wiskerke, J.S.C., Urban Food Systems, pp. 1–25, 2015.
[5] Ramachandra, Nair, P.K., *An Introduction to Agroforestry*, Springer Netherlands, XIV, p. 499, 1993.
[6] Reyburn, S., Évaluation de la contribution de l'agriculture urbaine communautaire montréalaise à l'amélioration du cadre de vie. Thèse. Québec, Université du Québec, Institut national de la recherche scientifique, Doctorat en études urbaines, p. 246, 2006.
[7] Montreal, Regroupement des éco-quartiers de Montreal. REQ. https://www.eco-quartiers.org/ (accessed on 3 April, 2018).
[8] Boudreau, J.-A., Hamel, P., Jouve, B. & Keil, R., Comparing metropolitan governance: The cases of Montreal and Toronto. *Progress in Planning*, **66(1)**, pp. 7–59, 2006.
[9] Beaudet, G., Un avenir pour les ruelles ?, Montréal campus, 2017. http://montrealcampus.ca/2017/10/un-avenir-pour-les-ruelles-vertes-2/ (accessed on 1 May, 2018).
[10] Guelper, N., Un avenir pour les ruelles ?, Montréal campus, 2017. http://montrealcampus.ca/2017/10/un-avenir-pour-les-ruelles-vertes-2/ (accessed on 1 May, 2018).
[11] Politique d'agriculture urbaine de la ville de Havane. Observatoire Villes Inclusives, p. 10, 2010.
[12] Gonzalez Novo, M. & Murphy, C., Urban agriculture in the city of Havana: a popular response to a crisis, No date.
[13] Howorth, C., Convery, I. & O'Keefe, P., Gardening to reduce hazard: urban agriculture in Tanzania. *Land Degradation and Development Journal,* **12(3)**, pp. 285–291, 2001. https://doi.org/10.1002/ldr.441
[14] Körösi, S., Nehogy az ördög ülve találjon, 2015. http://www.film-documentaire.fr/4DACTION/w_fiche_film/43920_1; https://vimeo.com/119547882
[15] Tenk A., et al., Csepel természeti képe [The physiography of Csepel], pp. 116, 88–89, 2014.
[16] BFVT Ltd. Green Infrastrucure Concept of Budapest, 2017. http://budapest.hu/Documents/V%C3%A1ros%C3%A9p%C3%ADt%C3%A9si%20F%C5%91oszt%C3%A1ly/II.%20k%C3%B6tet%20-%20Koncepci%C3%B3.pdf
[17] Faczanyi, S., Budapesti közösségi kertek tájépítészeti és szociológiai értelmezése, Szent Istvan Egyetem, 2017.
[18] Burky, J., Agroforestry, a decade of development, Chapter 16. *Exploitation of the Potential of Multipurpose Trees and Shrubs in Agroforestry*, p. 345, eds. H.A. Steppler & P.K. Ramachandran Nair, ICRAF, 1987.
[19] Agroforestry systems in China. Editors Zhu Zhaohua, Cai Mantang, Wang Shiji and Jiang Youxu. Published Jointly by: Chinese Academy of Forestry, People's Republic of China and International Development Research Centre, Canada. IDRC CRDI, p. 227, 1991.
[20] Gliessman, R.S., *Agroecology: The Ecology of Sustainable Food Systems*, 2nd edn., CRC Press, Taylor & Francis Group, p. 384, 2007.
[21] Bourguignon, C., Bois Raméal Fragmenté, BRF, LAMS, 2008. http://www.lams-21.com/artc/LAMS/1/fr/ (accessed on 3 May, 2018).
[22] Association Française d'Agroforesterie, AFAF, Agroforesterie et viticulture. http://www.agroforesterie.fr/documents/fiches-thematiques/fiche-thematique-AFAF-principes-agroforesterie-vigne-Viticulture.pdf
[23] Becquey, Boisement mélangé, Forêt-entreprise n°209-mars 2013.

[24] United Nations University. UNU. http://www.fao.org/docrep/s1930e/s1930e02.htm; http://archive.unu.edu/unupress/unupbooks/80824e/80824E0h.htm#7%20Pacific%20Island%20urban%20agroforestry (accessed on 1 October, 2017).
[25] United Nations Development Program, UNDP, 2012. http://www.undp.org/content/undp/en/home.html (accessed on 3 October, 2017).
[26] Crawford, M., *Creating a Forest Garden: Working with Nature to Grow Edible Crops*, UIT Cambridge Ltd, p. 384, 2010.
[27] Mann, S., et al., *Urban Agroforestry: Connecting Agroecology, Permaculture, Urban Forestry and Urban Agriculture into Urban Food Forests*, Center for Agroforestry, University of Missouri. No date.
[28] Spiegelaar, N., Tsuji, L. & Oelbermann, M., The potential use of agroforestry community gardens as a sustainable import-substitution strategy for enhancing food security in subarctic Ontario, Canada. *Sustainability*, **5(9)**, pp. 4057–4075, 2013. https://doi.org/10.3390/su5094057
[29] AGFORWARD, Hungary, Vityi, A., no date. http://www.agforward.eu/index.php/en/wood-pasture-in-hungary.html (accessed on 3 June, 2016).
[30] European Environment Agency, EEA, 2007. https://www.eea.europa.eu/ (accessed on 5 May, 2018).
[31] Homeless statistics, https://homelessworldcup.org/homelessness-statistics/ (accessed on 10 June, 2018).
[32] Szaller, V., Szabó, V., Diószegi, M.S., Magyar, L. & Hrotkó, K., *Urban Alley Trees in Budapest*, Slovak University of Agriculture in Nitra, pp. 28–31, 2014.
[33] Bouton, S., Je suis le Sorbier. Edition Actes Sud. Collection Je suis l'arbre, p. 25, 2008.

MAPPING THE AGRODIVERSITY IN BOGOTÁ – THE PLATFORM MAPEO AGROECOBOGOTÁ

K. PINILLA, B. HOINLE, A. MAHECHA-GROOT & J. CEPEDA
Agustín Codazzi Geographical Institute (IGAC), University of Hamburg, Rutgers University, National University of Colombia.

ABSTRACT

Counter-mapping is a useful tool to counteract the hegemonic forms of creating maps and overcoming territorial logics of domination. The purpose of counter-mapping projects is mapping and visualizing information to be monitored in real time and generate collaborative creation processes for the creation of alternative learning networks and interactive community cartography that works as a territorial appropriation tool. In Colombia, the counter-mapping platform AgroEcoBogotá is inspired by the social, environmental and agroecological movements that create new ways of doing and being in this world through agroecological initiatives, allowing the emergence of transformative pathways of the agrifood system in the metropolitan area of Bogotá, the capital of the country. The platform is a tool to promote and give visibility to the movement of agroecology in Bogotá, through the mapping of *huertas urbanas* (urban farms and gardens) and other urban agriculture projects. The aim is to create solidarity links between different urban and peri-urban food initiatives of Bogotá and its neighborhoods and to connect the existing networks of urban agriculture with the rural sphere. The motivation to create the platform was the observation and acknowledgment of the lack of a network of networks that could encourage the interaction and strengthening of agroecological initiatives. With the example of one mapping-point, *Casa Taller Las Moyas* in the Eastern Hills of Bogotá (*Cerros Orientales*) we want to illustrate the importance of *huertas urbanas* as spaces of local community empowerment. In this context, counter-mapping also serves as a resource for resistance and grassroots appropriation of space. As a conclusion, we will reflect about the challenges that *AgroEcoBogotá* is facing in the effort to make visible the agroecological movement of Bogotá, as well as, to strengthen the existing networks in Bogotá and their rural food supply connections in different parts of the region of Cundinamarca.
Keywords: agroecology, counter-mapping, critical cartography, social movements, urban agriculture.

1 INTRODUCTION

Recently, the urban agriculture movement in Bogotá, the capital of Colombia, is growing and is attracting public attention. During the last decades, the metropolitan region of Bogotá was growing at an accelerated tempo due to the arriving of a considerable number of migrants from different rural areas, many of them displaced by the dynamics of the armed conflict. Therefore, the huertas urbanas, hereafter referred to as urban gardens, have become spaces which firstly offer means for subsistence but in the long run can contribute to community empowerment, especially in the lower class neighborhoods. The huertas are fruit and vegetable gardens, but in the context of Colombia la huerta is also a privileged space in the life of rural communities. Generally, it is located next to the houses, and farmers weave a vital relationship with the earth and territory, through the diversity of plants and microorganisms in the ground. In an urban context, las huertas urbanas are also spaces for growing food, as well as, spaces for other non-productive functions such as education, food sovereignty, social cohesion. Despite the enormous potential of urban agriculture to integrate rural migrants in urban contexts such as Bogotá, efforts are still required to be recognized as a possibility for inhabiting the city in a more decent way.

In this context, the initiative *AgroEcoBogotá* (https://agroecobogota.crowdmap.com/) was launched as a platform (Fig. 1) [1] to give greater visibility to the existing urban farming

DOI: 10.2495/DNE-V13-N4-407-414

projects. Its aim is to deepen and strengthen the links between the various agroecological projects in the different parts of the city and to promote the local agroecological movement. The participation within the platform allows open access to the local urban farmers who can directly join the mapping process. Through its interactive configuration the platform gives the possibility to georeference information in real-time from every place of the world. This allows to generate processes of collaborative cartographic creation, both digital or analogue. Therefore, the initiative opens up the way for the exchange of knowledge about the reproduction of native seeds, medicinal plants and preparation of local recipes. Moreover, it aims to promote the food networks with the regional agroecological projects in the department of Cundinamarca (Colombia is divided into 32 *departamentos* (Departments), which are in turn subdivided into *municipios* (municipalities) and cities). By doing so, the initiative *AgroEcoBogotá* tries to challenge the hegemonic forms of knowledge production. This means forms of knowledge production that were perpetuated by dominant sectors of the society where the economic power (e.g. agribusiness) is situated. Instead of reproducing dominant representations of the urban landscape the methodology of counter-mapping opens up spaces for the visions, perspectives, and experiences of the local inhabitants. So, the intention is to give greater public visibility to groups who have been affected by territorial conflicts which often are not visible in the conventional maps. This kind of collective mappings could be called 'a process of creation which subverts the place of enunciation to challenge the dominant narratives about territories' [2]. Actually, the mapping process is constituted by 80 references of urban and periurban agroecological projects in Bogotá and its surroundings (Fig. 1). Nevertheless, in the metropolitan area of Bogotá there are still much more agroecological initiatives to map and grassroots processes to visualize.

Figure 1: Homepage from AgroEcoBogota platform.
Source: Taken from agroecobogota.crowdmap.com [1].

This paper has five sections. In the first section, we give an overview of the relation of social mapping and real-time mapping, and its importance of empowering citizen participation for the identification of socio-spatial change and transformative actions in space. Following that, we reflect on the relevance of applying collective mapping processes in the context of urban agriculture. In the third section, we analyze how the proposal of counter-mapping was put into practice in the creation of the platform *AgroEcoBogotá* by mapping the initiatives of urban agroecology in the metropolitan region of Bogotá. In the fourth section by highlighting one mapping point – the agroecological project *Casa Taller Las Moyas* – in the Eastern Hills of Bogotá (*Cerros Orientales*) we want to explore the potential of the mapping tool for processes of appropriation of space. *Casa Taller Las Moyas* is located in a context of ecological degradation and social marginalization in the informal neighborhood San Luis, so it constitutes a place of territorial conflicts on one hand and local communitarian processes of resistance on the other hand which makes this example especially interesting in the context of counter-mapping. Finally, in the fifth section, we will discuss the opportunities and challenges that the platform offers as a virtual space for promoting the visibility of the agroecological networks in the metropolitan area of Bogotá. As the mapping process of *AgroEcoBogotá* (Fig. 1) [1] is still in process (as most social processes are), we do not finish the article with final results, instead we conclude with an outlook about the perspectives, potentials and challenges for the future.

2 FROM SOCIAL MAPPING TO REAL-TIME MAPPING

Within the dominant cartographic representations, an entity is responsible for writing and marking the territory in a unified way, denying the possibility of seeing the different ways of perceiving and interacting in space. Instead, social cartography tries to make visible and legitimize the struggle for the search for alternative representation spaces of the different social actors involved in a specific territorial dynamics [3], [4]. Citizen participation online is an instrument that promotes the generation of collective knowledge and interactively allows to identify alternatives that help understand and act against urgent territorial issues. Social mapping has become an enhancing device for citizen participation for the identification of socio-spatial change and transformation actions. In this regard, Herrera [5] states

> "[social-mapping] emerges as a new tool for planning and social transformation. It is currently being used as support in community organization processes through decentralized and democratic participatory planning in which the participation of all local actors acquires great importance. Besides, the social cartography of a territory is proposed as a new instrument for the construction of knowledge from participation and social commitment, enabling the transformation of it." [5]

The emancipatory ideals of social-mapping, together with the potential of free software, have created counter-mapping tools such as crowdmapping. Ushahidi, the technological corporation, is an excellent example of an open-access and open-source real-time mapping portal, that have the possibility of personalizing themselves through the collective knowledge of citizenship. Due to the difficulty in obtaining accurate information about the detail of the civil rights violations that were co-occurring in some African cities, Ushahidi arose as an alternative for mapping to more real story told by the citizenship. In short, the contribution of the same citizens was the most legitimate knowledge of the reality that they were living. Citizen participation was the most valuable element of denunciation and social control that legitimized the birth of geographic information system to systematize collective knowledge generated in real time [6].

We emphasize that for the creation of these platforms there are two different methods, which go according to the public that needs them. For the developers, the free open source database (ODbL) was made available where a programming language is used (either for Windows or Linux) with the possibility of collective improvement, and, while retaining the copyright of the generators of the information, that is, facilitating the creation of collective knowledge on the web and free licensing. For people who do not master programming words, the extension releases the standard code base that allows the automatic creation of platforms, accompanied by a broad level of web customization [6].

While there is a consumer relationship regarding access to technologies and the Internet in most Latin American countries, this does not mean that valuable and innovative knowledge cannot be generated to understand territorial dynamics. What it needs is to make the territorial actors visible and create interaction spaces of their knowledge intending on creating a shared and emancipatory knowledge.

3 WHY MAPPING URBAN AGRICULTURE?

Generally, since the formation of large cities due to industrialization, urban agriculture has been developed in cities to alleviate food shortage and improve the quality of life. With great boom periods related to crisis, such as the world wars, and now, environmental and social problems that have led to think of urban agriculture as an alternative for countries in both the North and the Global South. Particularly in the Global South, urban agriculture has been studied from its role in the construction of livelihood strategies for people who do not have access to social safety nets [7–9].

Urban agriculture in Bogotá response to the global trend we mentioned above, related with urban expansion and local migrations during the first half of twenty century. This local migrations were mostly from the countryside, principally from the departments of Boyacá, Cundinamarca, Santander and Tolima [10]. Thus, we can imagine that these peasants bought not only cultivation habits and their own knowledge, but also native seeds and recipes, which were resignified in the urban space. The orchards in Bogotá are therefore spaces for exchange of knowledge and urban–rural practices.

The institutionalization of urban agriculture in Bogotá takes place in 2004, during the mayor's office of Luis Eduardo Garzón (2004–2007) in the framework of food security policy. Currently, in Bogotá there is public policy on urban agriculture or 'Agreement 605 of 2015 of the Council of Bogotá', however, this strategy has been limited implementation due to lack of budget. One of the proposals of this policy is the SIGAUPA or geographic information system of urban and peri-urban agriculture, tool that is under construction due precisely to budget constraints.

In a global regarding of urban agriculture, it can be considered that its practice has been continuous, but in many cases it is found that the state supports in contrast have been intermittent, so the presence of civil society around the urban farming is vital to their survival and it remains important today throughout the world [11]. This is also applicable to the decision making tools for the development of this activity, thus, the proposal of *AgroEcoBogotá* responds to the need to generate this type of self-managed and independent strategies.

Based on the big picture, we consider that the mapping of urban agriculture will allow us to understand:

1. The scale of where and what services are available in the city to make possible its monitoring, the close follow-up of the implementation of public policies on urban agriculture,

the management of community-applied supplies, that is, this vision proposes to mapping as tool for decision making.

2. The map as a tool for the communication of untold and alternative stories of doing agriculture, as well as for the transmission of important information within the construction of territoriality.
3. Support for assessing the sustainability of the city and its degree of food sovereignty from the base communities.
4. Visibility and empowerment of people and groups dedicated to urban agriculture.
5. Support for better management of local networks and the district network of urban farmers.
6. Show the current and potential importance of urban agriculture for local and regional development.

4 THE PROCESS OF THE PLATFORM MAPEO AGROECOBOGOTÁ

The purpose to found the platform *AgroEcoBogotá* was to create networks between the different urban and periurban gardens of Bogotá. The idea came up because of three observations: (a) In many cases the urban farmers of the same neighborhood don't know each other; (b) More and more inhabitants of the urban neighborhoods would like to buy fresh organic vegetables but they do not know where; (c) In Bogotá, there are existing several networks at the level of the local neighborhoods and districts. But at that time there wasn't any network at the metropolitan scale that connects all the local networks.

These observations were made by people engaged in fieldwork in the peri-urban and urban gardens of Bogotá. They gathered for an exchange about ideas about their universitary theses and social activism. In these meetings the idea came up to visualize all the urban farming projects they came to know during their field visits. With the idea to strengthen the urban farming movement of Bogotá they started to organize the platform since 2014. The first announcement was made by the 'I Encuentro de Agricultura y Educación para la permanencia de la vida – Permaculturas Colombia' (First Meeting of Agriculture and Education for the Sustainability of Live – Permacultures in Colombia) organized by the Department of Geography at the National University of Colombia. The first rapprochement of mutual learning with the technology of the platform was pursued with the network for alternative education REEVO (*Red de Educación Alternativa*). This organization already had undertaken a mapping process with initiatives of all over the world.

Another important precedent was a audiovisual project about urban agriculture in Bogotá for the online journal *Razón Pública* initiated by a group of scholars and activists. The collective was visiting several urban gardens in the central and peripheral lower class neighborhoods like La Perseverancia, Diana Turbay (local district Rafael Uribe Uribe) and in the rural side of Usme (still part of Bogotá). During the shooting of the videos and the interviews with the local urban farming activists we recognized the huge diversity of the urban agriculture landscape of Bogotá: Every urban garden has its own characteristics. Some work more with native seeds, others with medicinal plants and some are engaged in the elaboration of traditional recipes.

The urban garden in the local neighborhood of Diana Turbay was an impressive example of how the different urban gardens are interconnected with each other at local scale. There are different forms of exchange that the urban gardeners are practising like exchange of recipes, seeds, plantlets and knowledges. Moreover, every month, they meet to help each other in the gardens, a practise called '*minga*' in the Andean Regions. So, by their daily practices, they are already experimenting economic alternatives and creating Commons in a material and symbolic way: the material foundations to build up a garden (soil, seeds) and symbolic (agroecological knowhow). By doing so, they are creating links of solidarity and neighborship within their district. In cases like this, the platform could help to facilitate the communication and communitarian organization between the urban gardens. One example for the creation of networks at the district scale is the Network of Communitarian Gardens in Ciudad Bolívar (Red de Huertas Comunitarias in Ciudad Bolívar). They are also organizing '*mingas*' and coordinating agroecological markets with the vegetables from the gardens in the lower class neighborhoods of Ciudad Bolivar. Nevertheless, in most of the neighborhoods, the urban garden initiatives are rather disconnected and the people with the same agroecological ideas don't know each other. For this reason we came up with the idea of the platform *Mapeo AgroEcoBogotá* to give the urban gardening projects more visibility and to create more networks on a wider scale between the already existing urban gardening networks. We were inspired by some international mapping processes like the mapping of the communitarian and intercultural gardens of Berlin undertaken by the Kollektiv Orangotango. In this mapping process they referenced 99 urban gardening projects [12].

With this proposal in mind we met at the Encounters of Permaculture held up near by the National University of Bogotá. In this time the collective of *AgroEcoBogotá* was founded, originally by four persons motivated by the interest to visualize the diversity of the urban agriculture of Bogotá in a cartographic way. The collective was formed by people of different areas and countries but all of them engaged in the issues of agroecology and participatory action research. Since 2015 we were meeting constantly to define the basic configurations of the platform. After several discussions we defined three categories to differentiate the distinct focal points of the urban gardens: seeds, alimentation and education. This differentiation is useful to promote the different objectives that the gardens are pursuing in the reproduction of native seeds and in spreading their experiences by workshops. We installed the platform in an open source format based on Open Street Map to guarantee the open access and interactive configuration of *AgroEcoBogotá*.

5 EXAMPLE OF ONE MAPPING POINT – *CASA TALLER LAS MOYAS* IN SAN LUIS, CHAPINERO, BOGOTÁ

In the eastern hills of Bogotá, in the informal neighborhood called San Luis, where urbanity meets rurality and the dark green forests of the mountains, there is the location of Casa Taller Las Moyas. This is a self-managed community project for the care of children in the absence of the state, and the urban agroecological initiative is part of the strategies of this vital project. This urban agroecological project is situated within an informal neighborhood called San Luis. In the streets of San Luis you find peasants with the traditional *ruanas* (typical Colombian poncho) and youth people listening to urban reggaeton music. San Luis is a self-built neighborhood, with the will and ingenuity of the villagers themselves; it was a transit point due to the transportation of coal and wood from Bogotá to La Calera. This place belongs to the borough of Chapinero and is populated by migrants from the countryside to the city or from the city to the surroundings due to the multiple forms of political and economic violence in Colombia. The people had to build up their own infrastructure, so until today the neighborhood has its own community-based aqueduct. The inhabitants were fighting for their formal recognition by the administration of Bogotá, but until now only one part of the neighborhood was legalized. Recently, San Luis is experiencing rapid transformation. As one part continues as an informal, marginalized neighborhood, in the other part real estate firms are investing in upper class housing projects. This creates more and more social inequality within the neighborhood. In this conflictive context, the urban agroecological project *Casa Taller Las Moyas* tries to build up an alternative of communitarian organization and integration of different generations.

Casa Taller is like an open meeting point in San Luis. Children come here to do their homework, especially when their parents are all day long at work and they would be alone at their house. Some young people meet to make music, soaps, creams, ointments, grow plants, and all the activities that the same relationships of life make them learn. They are considered an active community school and with the task of rethinking as worthy human beings on a constant basis. On weekends they plan activities such as the mingas and people from different parts of Bogotá join the work in the urban garden and learn more about plants and ancestral recipes. *Mingas* are an Andean tradition of collaborative work that was practiced since generations by the indigenous communities. The aim is to organize work in a solidarity form based on mutual self-help. These ancestral practices find also a revival in the today's urban farming projects. To the mingas in *Casa Taller Las Moyas* usually come activists of the network 'Guardianes de Semillas' (Guardians of Seed). These women share their knowledge about seed reproduction and conservation in urban gardens. Therefore, the urban garden in *Casa Taller Las Moyas* is a place of exchange between different, experience-based ancestral knowledges (diálogo de Saberes). By doing so, the women of the seed movement contribute to maintain the hugh agrobiodiversity in seeds (maize, beans, potatoes, herbs) that is circulating between the urban farming projects in Bogotá. For Nicolasa, the founder of the project, Casa Taller is a free education initiative in which sowing is a pedagogical tool to reconnect with nature within the city. Casa Taller is a family of generations of the city of Bogotá that meet to learn and share in a collaborative and emancipatory way, they offer and demand communitarian learning for the production of different personal and useful things for life – deodorants, creams, ointments, notebooks among others. For the children, the things that they elaborate are always a possibility for learning from a young age how they can generate their own money and dignify themselves from the pedagogy of love. This pedagogy allows to find the power that all human beings have to be and feel worthy thanks to the responsibilities they assume, instead of the begging discourse. At the time of creating the products, children and young people apply the knowledge related to the different branches of knowledge, thus applying project-based learning. They have received since the foundation of Casa Taller collaborative stimuli from the same inhabitants and the network of contacts that they have achieved thanks to their work and effort. Casa taller is part of the *AgroEcoBogotá* mapping for being a reference for agroecological movement concentration.

6 CONCLUSION: CHALLENGES AND PERSPECTIVES

The mapping platform *AgroEcoBogotá* has still some challenges to face. One of them is to create more opportunities for an encounter between the agroecological initiatives and the inhabitants of Bogotá where they could exchange more their work and knowledge. There are still some challenges to integrating more rural migrants and displaced people in the urban gardens as an alternative for communitarian organization in their local neighborhoods. Another challenge is to transcend the registrations already made from the sum of several individual points to a collective sense of being part of a pivotal social movement. The latter would also be useful to multiplicate collaborative ethics within the urban gardens based on the principles of agroecology and the commons.

The mapping platform has a enormous potential to strengthen the agroecological movement in Bogotá. The platform could be a tool to link more the rural agroecological producers and urban consumers and therefore create spaces for experiences with community supported agriculture with the idea to construct a juster and more sustainable food system between city and countryside.

Furthermore, the platform has also a significant potential for the documentation and research in peri-urban and urban agriculture. As the information gets continuously actualized, the platform is much more dynamic and close to the urban farming activists. However, the most significant aspect is that the urban farmers themselves are appropriating the platform and use it for their purposes. How can we promote the autonomy of the mapping process? Also, which tools would be helpful for this aim? These are some questions that guide the following processes with the platform in the future.

REFERENCES

[1] Agroecobogotá. Crowdmapping extension from Ushahidi Technology. Online. https://agroecobogota.crowdmap.com (accessed on 23 May, 2018).

[2] Iconoclasistas (eds.). *Manual del mapeo colectivo: recursos cartográficos críticos para procesos territoriales de creación colaborativa*. Tinta Limón: Buenos Aires, 2013.

[3] Barrera Lobatón, S., Reflexiones sobre Sistemas de Información Geográfica Participativos (SIGP) y cartografía social. *Cuadernos de Geografía-Revista Colombiana de Geografía*, **(18)**, pp. 9–23, 2009. https://doi.org/10.15446/rcdg.n18.12798

[4] Oslender, U., Ontología relacional y cartografía social:¿ hacia un contra-mapeo emancipador, o ilusión contra-hegemónica? *Tabula Rasa*, **(26)**, pp. 247–262, 2017.

[5] Herrera, J. Cartografía social. Online. https://juanherrera.files.wordpress. com/2008/01/cartografia-social.pdf (accessed on 23 May, 2018).

[6] Okolloh, O., Ushahidi, or 'testimony': Web 2.0 tools for crowdsourcing crisis information. *Participatory Learning and Action*, **59(1)**, pp. 65–70, 2009.

[7] Smit, J., Nasr, J. & Ratta, A., *Urban Agriculture: Food, Jobs and Sustainable Cities*. UNDP: New York, 1996.

[8] Altieri, M.A., Hecht, S., Liebman, M., Magdoff, F., Norgaard, R. & Sikor, T.O., *Agroecología: Bases Científicas Para Una Agricultura Sustentable*. Nordan-Comunidad, 1999.

[9] Lynch, K., Binns, T. & Olofin, E., Urban agriculture under threat: The land security question in Kano, Nigeria. *Cities*, **18(3)**, pp. 159–171, 2001.

[10] Rodriguez, D.R. & León, T.E., *Agricultura urbana en Bogotá: adaptación cultural a los ecosistemas*. En: Nail, S. Alimentar la ciudades. Ed. Universidad Externado de Colombia, pp.181–210, 2018. ISBN 978-958-772-927-6.

[11] Moran Alonso, N., Agricultura urbana: un aporte a la rehabilitación integral. Papeles de relaciones ecosociales y cambio global, 111 pp. 99–111, 2010 Online. http://oa.upm.es/12160/1/INVE_MEM_2010_76416.pdf (accessed on 23 May, 2018).

[12] Orangotango. Kollectiv für kritische bildung und kreativem protest. Online. http://orangotango.info/projekte/kollektives-kartieren/kritische-karten-2/ (accessed on 23 May, 2018).

THE FATE OF COMPOST-DERIVED PHOSPHORUS IN URBAN GARDENS

G. SMALL[1], P. SHRESTHA & A. KAY
Department of Biology, University of Saint Thomas, United States.

ABSTRACT

Urban agriculture has been expanding rapidly in recent years, and it has the potential to recycle nutrients from local food wastes into new foods through the use of compost as growth medium. Composts typically have low nitrogen: phosphorus (N:P) ratios, and in urban gardens, when composts are typically applied to soils annually to meet the N demand of crops, excess P can build up and be prone to leaching. We measured dissolved P (PO_4^{3-}) and N (NO_3^- & NH_4^+) losses in the leachate from experimental raised-bed garden plots that received one of two commonly used composts (municipal organics compost derived from food scraps or cow manure derived compost) at three different application levels (15%, 35%, or 50% by volume). PO_4^{3-} concentrations in leachate from garden plots receiving manure composts were high, ranging from 5 to 11 mg/L, depending on the application level. Leachate PO_4^{3-} concentrations from plots receiving municipal organics composts were an order of magnitude lower, ranging from 0.5–1.2 mg/L, while leachate PO_4^{3-} in garden plots receiving no compost was 0.3 mg L^{-1}. Cumulative mass of PO_4^{3-} lost through leachate during the growing season ranged from 1.2 to 4.2 g/m^2 for manure compost treatments, compared to 0.12–0.72 g/m^2 for municipal compost treatments, and 0.06 g/m^2 for soil with no compost. Leachate accounted for 0–37% and 18–45% of dissolved P and N exported from garden plots, respectively. The high application rate of municipal compost significantly increased crop yield relative to the control treatment. P lost from leachate and removed through harvest only represented 1–10% of total P applied as compost, suggesting that soil build-up was the dominant fate of P in this study. Our results illustrate the potential trade-off in urban agriculture between crop production and recycling P efficiently from urban food waste.

Keywords: compost, nitrogen, nutrient leaching, nutrient recycling efficiency, phosphorus, urban agriculture, water quality.

1 INTRODUCTION

Urban ecosystems are characterized as being extremely heterotrophic, with a high throughput of materials, and low internal cycling [1]. Cities rely on a land area 100–300 times larger than the city itself for the provisioning of food, water, and energy, and the assimilation of waste [2]. Urban agriculture has been proposed as a means of partially closing this loop, recycling nutrients from organic waste back into the human food system [3], and making cities less reliant on imported food [4].

Although there are many potential social, economic, environmental, and public health benefits of urban agriculture [5], [6], there is a potential tradeoff between maximizing nutrient recycling in cities and minimizing subsequent nutrient pollution through runoff or leachate. Organic wastes typically have low nitrogen:phosphorus (N:P) ratios relative to crop nutrient demand [7], so that if compost is applied based on crop N-demand, soils receive excess P. While soils have the capacity to naturally bind P, at high P concentrations this capacity can be exceeded, [8], and instead, soils become a source of P export. Over time, excess P applications can lead to P buildup along hydrologic flowpaths which contributes P losses to drainage water resulting in water pollution that lasts for decades or centuries [9].

1 *ORCHID: http://orchid.org/0000-0002-9018-7555*

DOI: 10.2495/DNE-V13-N4-415-422

Evidence from recent studies indicates the potential for excessive P application in urban agriculture. A study of urban food gardens in Chicago documented excessive fertilization and soil P levels averaged 263 ppm [10], ten-fold higher than optimal levels for vegetable gardens [11]. Thirty-eight percent of domestic gardens surveyed in Flanders had P levels classified as 'high' or 'very high' relative to optimal agronomic standards [12]. An analysis of Montreal's urban agriculture system [13] documented that 27 times more P is applied as compost than is harvested in crops each year, resulting in P loss or soil build-up of 175 kg P/ha/y across the 18 km^2 of cultivated land in the city. An analysis of urban and peri-urban farms in Nigeria, Burkina Faso, and Mali documented large annual surpluses of P (83–780 kg/ha/y) and N (74–1127 kg/ha/y) resulting in nutrient loss to groundwater [14]. Given the rapid expansion of urban agriculture, this excess P has the potential to create hotspots of nutrient pollution.

Despite increasing evidence of over-application of P in urban agriculture, we lack understanding of how different soil amendment practices affect P use efficiency, and what fraction of this excess P is lost or exported through leachate. We measured leachate export and nutrient recycling efficiencies for P and N for vegetables grown in replicated raised bed plots using two different types of compost at three different compost: soil ratios. We hypothesize that nutrient use efficiency decreases and nutrient export through leachate increases with increasing compost application rates.

2 METHODS

2.1 Experimental set-up

This experiment took place at the University of Saint Thomas (Saint Paul, Minnesota, USA), using 36 replicated raised-bed garden plots with dimensions of 1 m length x 0.3 m width x 0.3 m depth. Prior to the experiment, soil in these plots (which had been used in previous years) was removed and homogenized. Initial soil chemistry, and nutrient content of the two compost varieties are presented in Table 1.

Table 1: Physiochemical properties of garden soil and compost samples.

	Garden soil	
NO_3–N (ppm)	3.5	
Bray P (ppm)	20	
Available K (ppm)	72	
Organic matter (%)	3	
pH	7.8	
	Municipal compost	**Manure compost**
Total C (%)	19.26	39.17
Total N (%)	1.11	1.68
Total P (%)	0.16	0.60
Total K (%)	0.41	0.60
C:N ratio (mass)	17.35	23.32
N:P ratio (mass)	6.94	2.80
Bulk density (g dry mass/L)	490.8	116.4

Each raised-bed plot was randomly allocated one of seven compost-soil mixtures (ratios are based on volume): 100% soil; 15% municipal compost + 85% soil; 35% municipal compost + 65% soil; 50% municipal compost + 50% soil; 15% manure compost + 85% soil; 35% manure compost + 65% soil; and 50% manure compost + 50% soil. The amount of P added in the form of compost ranged from 11 to 36 g in the municipal compost treatments, and from 9 to 32 g in the manure compost treatments. The amount of N added in the form of compost ranged from 76 to 248 g in the municipal compost treatments, and from 28 to 91 g in the manure compost treatments. The 100% soil treatment was applied to six replicate plots; all other treatments were applied to five replicate plots.

One row of radish seeds (Easter Egg Radish, small tricolored, from Johnny's Selected Seeds) and two rows of carrot seeds (a mixture of Yellowbunch F1 Carrot, Purple Haze F1 Carrot, and Nelson F1 Carrot, from Johnny's Selected Seeds) were planted in each raised-bed plot on 20 May 2016. After germination, seedlings were thinned to a density of approximately one seedling/2.5 cm.

Plots were watered every two to three days throughout the experiment, and soil moisture was generally maintained between 15 and 20%. Water was distributed evenly over each plot for a set amount of time (typically 30, 45, or 60 seconds), and volume of water was estimated by measuring the time required to fill an 11-L bucket.

2.2 Plant analysis, and nutrient use efficiency

Radishes were harvested on 29 June, and replanted on 6 July. Carrots and replanted radishes were harvested on 15 August, marking the end of the experiment. The total experiment ran for 88 days. At each harvest, total wet mass of plant material removed from each plot was recorded, with stems and leaves measured separately from edible root biomass. A subsample of plant material from each experimental treatment was weighed, dried at 60°C for 72 hours, and re-weighed. These values were used to calculate dry weight:wet weight ratios for each crop, and these subsamples were then processed for total carbon (C), N, and P content. Dried plant samples from each subplot were ground into fine powder (<0.5 mm). For total C and N, samples were analyzed in duplicates by combustion method in a Thermo Electron Flash EA 112 Series CN elemental analyzer (Thermo Fisher Scientific, Waltham, MA, USA). Total P was measured by ashing samples (500°C for 2 h) followed by persulfate digestion, and analyzed spectrophotometrically using ascorbic acid [15], using NIST apply leaf standard (NIST SRM 1515, Sigma-Aldrich, St. Louis, MO, USA) as a reference.

Nitrogen use efficiency (NUE) and phosphorus use efficiency (PUE) were calculated as the amount of N or P removed in crops relative to the total N or P added as compost to each plot (these terms were not calculated for control treatments where no compost was added). Mass of N or P removed in crops was calculated by multiplying percent nutrient content of the crop by dry biomass.

2.3 Nutrient leachate measurement

We collected leachate by installing a lysimeter in the center of each raised bed. Lysimeters consisted of a plastic funnel with a diameter of 23 cm, attached to a 1 L narrow-mouth polyethylene bottle. Wire mesh covered the hole in the funnel, and the funnel was filled with a 3 cm layer of pea gravel. A 0.6 m segment of Tygon tubing extended from the bottom of the bottle to the surface. Lysimeters were buried so that the top of the funnel was level with the base of the raised-bed, below the 0.3 m layer of soil-compost mixture. Lysimeters were sampled every 3–4 days throughout the growing season by using a 50 mL syringe to empty the collector bottle. The volume of leachate was recorded, and a 15 mL subsample was transferred to a plastic scintillation vial for subsequent analysis of leachate PO_4^{3}–, NO_3–, and NH_4^+ concentrations. Leachate concentrations were measured from all lysimeters for which >5 mL of leachate was recovered. Water samples were either analyzed immediately, refrigerated if run within 48 hours, or frozen if analyzed later. Samples were diluted at a 1:10 ratio with deionized/distilled water. Leachate PO_4^{3}– concentrations were analyzed using Hanna Instruments Phosphate Low Range Portable Photometer (HI96713). Leachate NO_3– and NH_4^+ concentrations were analyzed using YSI Professional Plus Multiparameter Meter fitted with a Pro Series 1006 nitrate ISE sensor and a Pro Series 1004 ammonium ISE sensor.

Leachate nutrient mass was calculated for individual sampling events (n = 22) by multiplying nutrient concentration of the subsample of lysimeter water by the total water volume in the lysimeter: nutrient mass (mg) = nutrient concentration (mg/L) × leachate volume (L). The resulting mass numbers were divided by the lysimeter funnel area (approximately 0.04 m^2) and scaled to the area of the raised-bed garden plot (0.3 m^2). Cumulative mass of nutrients exported from the garden plots over the growing season was determined by summing all the individual mass flows.

We used an Analysis of Covariance to test for the effect of compost N or P added and compost type (manure or municipal) on crop yield, crop %P and %N, PUE or NUE (phosphorus or nitrogen use efficiency, respectively), and inorganic P (PO_4^{3}–) or N (NO_3– & NH_4^+) leachate. Statistical analysis was conducted using JMP Pro 13.

3 RESULTS

Total harvested crop wet mass ranged from 1566 to 3593 g, and was positively related to the volume of compost added ($P = 0.0254$), but not compost type ($P = 0.4536$) or compost type × compost volume interaction ($P = 0.6048$).

Radish P-content ranged from 0.22 to 0.37% P, and was not affected by compost P added or compost type. Radish N-content ranged from 2.87 to 3.86% N (by dry mass), and was not related to compost N added ($P = 0.1344$), but was significantly higher for municipal compost treatments compared to manure compost treatments ($P = 0.0231$). Radish stems and leaves made up 33.7% of the total harvested biomass of radishes. Carrot P content ranged from 0.13 to 0.46% P, and was not related to compost P added or compost type. Carrot N-content ranged from 0.96 to 2.12% N, and was not related to compost N added or compost type. Carrot stems and leaves made up 53.8% of the total harvested biomass of carrots. Total harvested plant biomass had an average N:P ratio (by mass) of 8.63, compared to N:P ratios of 6.88 and 2.83 for the municipal and manure compost.

The mass of P removed in the form of crop biomass ranged from 0.66 to 1.83 g P. Total P harvested was not related to compost P added ($P = 0.85$) but was higher for municipal compost ($P = 0.0404$) with a significant compost P added × compost type interaction term ($P = 0.0205$) (Fig. 1A). The mass of N removed in the form of crop biomass ranged from 5.33 to 14.84 g N. Total N harvested increased with compost N added ($P = 0.0044$) and was higher for manure compost ($P = 0.0002$), with a significant compost N added × compost type interaction term ($P = 0.0073$) (Fig. 1B). Control plots (in which no compost was added) had a mean crop yield of 0.98 g P and 8.27 g N.

PUE ranged from 2 from 17% across plots. PUE was inversely related to compost P added ($P < 0.0001$). There was no effect of compost type ($P = 0.1202$), but there was a significant compost P added × compost type interaction term ($P = 0.0499$) (Fig. 2A). NUE ranged from 3 to 42% across plots. NUE was inversely related to compost N added ($P < 0.0001$), and was higher for municipal compost ($P = 0.0031$), with a significant compost N added × compost type interaction term ($P < 0.0001$) (Fig. 2B).

Cumulative PO_4^{3-} leachate ranged from 0.00 to 0.95 g P across plots. Cumulative PO_4^{3-} leachate increased with compost P added ($P < 0.0001$) and was higher for manure compost ($P < 0.0001$), with a significant compost P added × compost type interaction term ($P = 0.0002$) (Fig. 3A). Cumulative inorganic N leachate ranged from 0.00 to 22.04 g N across plots.

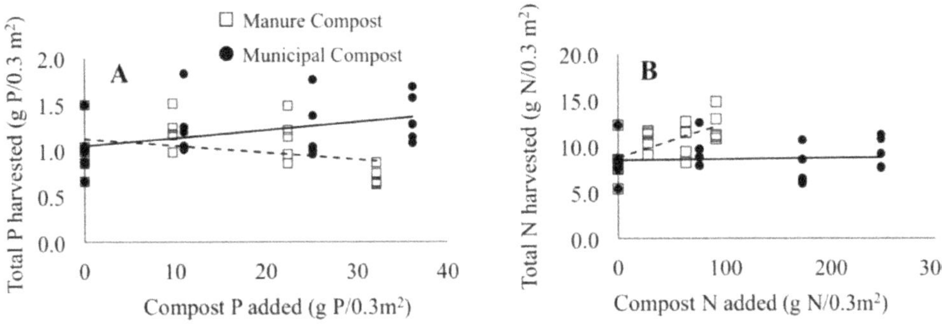

Figure 1: (A) Total P harvested versus compost P added for manure and municipal composts. (B) Total N harvested versus compost N added for manure and municipal composts.

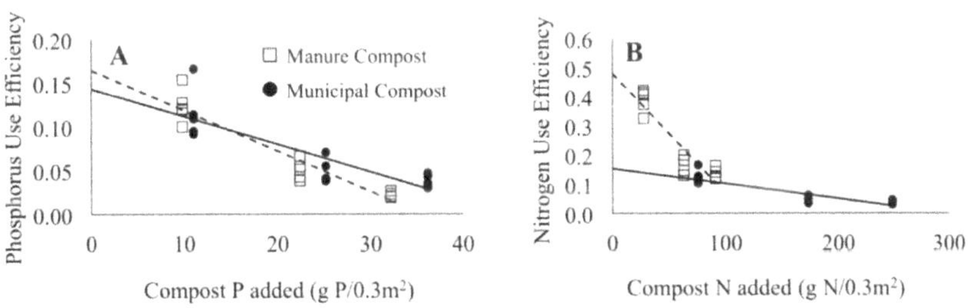

Figure 2: (A) Phosphorus use efficiency versus compost P added for manure and municipal composts. (B) Nitrogen use efficiency (defined as total N harvested relative to compost N added) versus compost N added for manure and municipal composts.

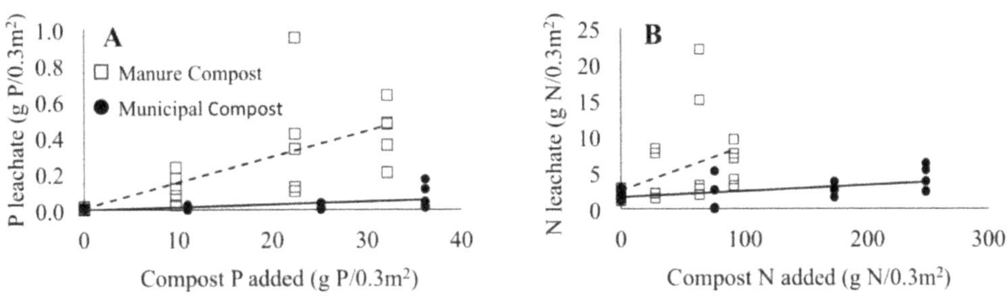

Figure 3: (A) PO_4^{3-} leachate versus compost P added for manure and municipal composts. (B) Inorganic N (NO_3^- & NH_4^+) leachate versus compost N added for manure and municipal composts.

Cumulative inorganic N leachate increased with compost N added (P = 0.0060) and was higher for manure compost (P = 0.0008), with a significant compost N added × compost type interaction term (P = 0.0325) (Fig. 3B). Leachate had a mean volume-weighted N:P mass ratio of 31.

4 DISCUSSION

Our results support the hypotheses that PUE and NUE decreased, and that P and N leachate increased, with increasing compost N and P application. The low N:P ratios of compost relative to crop biomass led to a greater application of P relative to N, but this imbalance was partially offset by the high N:P ratio of leachate.

The positive relationship between crop yield and compost application is indicative of a tradeoff between production and nutrient use efficiency that has been documented in conventional agriculture [16]. The low nitrogen and phosphorus use efficiency values for the higher levels of compost application were consistent with results previously reported [17] for a 50:50 mixture (by volume) of different food-waste derived composts and soil, in which PUE ranged from 2 to 8% and NUE ranged from 3 to 17%. These results suggest that efforts to recycle urban waste into urban food production may create additional negative externalities.

While compost treatments in this study applied nutrients at a level of up to 25-fold greater than crop N demand and 36-fold greater than crop P demand, this rate of compost application is consistent with common practices in urban agriculture [13]. Because compost is a slow-release fertilizer, with only a fraction of the total N and P becoming available to plants during the first year it is applied [18], [19], there may be a tendency among farmers and gardeners to over-apply compost if compost is applied annually based on anticipated crop N demand (i.e. without accounting for mineralization of older compost in the soil). Gardeners and urban farmers may apply large amounts of compost with goals of building soil organic matter and improving moisture retention. Our results show that low nutrient use efficiency and nutrient loss through leachate is a consequence of over-application of compost, not only for mobile nutrients such as N but also for P. Manure compost is widely used in urban and peri-urban agriculture [20], but our results suggest that this soil amendment may be particularly prone to leaching P and N, perhaps due to higher mineralization rates of labile organic matter, compounded by increased infiltration as a result of a very low bulk density (Table 1). The low N:P ratio of manure compost (ratio of 2.8; Table 1) also leads to a high rate of P build-up in the soil over time. One potential solutions may be to mix manure composts with feedstocks derived from woody wastes (i.e. woodchips, sawdust, woody mulch) that contain low P content and a higher C:N ratio. Woody wastes contain recalcitrant forms of nutrients that break down slowly and limit the leaching of P and N, and they can be added to a compost mix with a low N:P ratio (i.e. manure compost) or low C:N ratio (i.e. municipal compost in this case) to increase the overall C:N ratio of the final product. Achieving greater C:N ratio in the soil increases C availability, which promotes microbial immobilization of nutrients. N and P mineralization is subsequently reduced, and so is their availability for leaching or transformation to gaseous N_2O [21]. Given the stoichiometry constraints of microbial nutrient use, a C:N ratio greater than 20 is recommended to promote immobilization of N as opposed to mineralization [22]. More research is needed in finding ways to make stoichiometric adjustments to compost blends used for urban agriculture to increase crop nutrient use efficiencies and decrease compost nutrient leaching while maintaining crop yields.

Our results showed lower N and P leachate fluxes for municipal compost (Fig. 3), but similar NUE and PUE for a given amount of compost N or P applied (Fig. 2), indicating that more of these nutrients were retained in the soil, presumably becoming available over subsequent years. With continued growth of urban agriculture in the United States [23] and worldwide [24] and increased efforts to divert organic wastes away from landfills as a climate change mitigation strategy and recycle within cities through composting, the application of composts on urban farms and gardens may be expected to increase. Our results indicate that, while this strategy recycles nutrients back into the human food system, it might also turn urban gardens into hotspots of pollution.

ACKNOWLEDGEMENTS

Zachary Beckman, Casey Clemenson, and Brandon Paulson contributed to field measurements and laboratory analyses. Catherine Grant provided support with plant propagation. This study was supported in part by a National Science Foundation CAREER award (award number 1651361) to G. E. Small.

REFERENCES

[1] Burger, J.R., Allen, C.D., Brown, J.H., Burnside, W.R., Davidson, A.D., Fristoe, T.S., Hamilton, M.J., Mercado-Silva, N., Nekola, J.C., Okie, J.G. & Zuo, W., The Macroecology of Sustainablility. *PLoS Biology* **10**, p. e1001345, 2012.

[2] Rees, W.E. & Wackernagel, M., Appropriated carrying capacity: Measuring the natural capital requirements of the human economy. *Investing in Natural Capital: Ecological Economics Approach to Sustainability*, ed. A. Jansson, M. Hammer, C. Folke & R. Costanza, Island Press: Washington, DC, pp. 362–390, 1994.

[3] Baker, L.A., Can urban P conservation help to prevent the brown devolution? *Chemosphere*, **84(6)**, pp. 779–784, 2011. https://doi.org/10.1016/j.chemosphere.2011.03.026

[4] Grewal, S.S. & Grewal, P.S., Can cities become self-reliant in food? *Cities* **29(1)**, pp. 1–11, 2012. https://doi.org/10.1016/j.cities.2011.06.003

[5] Brown, K. & Jameton, A., Public health implications of urban agriculture. *Journal of Public Health Policy*, **21(1)**, pp. 20–39, 2000. https://doi.org/10.2307/3343472

[6] Smit, J. & Nasr, J., Urban agriculture for sustainable cities: using wastes and idle land and water bodies as resources. *Environment and Urbanization*, **4(2)**, pp. 141–152, 1992. https://doi.org/10.1177/095624789200400214

[7] Kleinman, P.J.A., Allen, A.L., Needelman, B.A., Sharpley, A.N., Vadas, P.A., Saporito, L.S., Folmar, G.J. & Bryant, R.B., Dynamics of phosphorus transfers from heavily manured coastal plain soils to drainage ditches. *Journal of Soil and Water Conservation*, **62**, pp. 225–235, 2007.

[8] Heckrath, G., Brookes, P.C., Poulton, P.R. & Goulding, K.W.T., Phosphorus leaching from soils containing different phosphorus concentrations in the Broadbalk experiment. *Journal of Environmental Quality*, **24(5)**, pp. 904–910, 1995. https://doi.org/10.2134/jeq1995.00472425002400050018x

[9] Sharpley, A., Jarvie, H.P., Buda A., May, L., Spears, B. & Kleinman, P., Phosphorus legacy: Overcoming the effects of past management practices to mitigate future water quality impairment. *Journal of Environmental Quality*, **42(5)**, pp. 1308–1326, 2014. https://doi.org/10.2134/jeq2013.03.0098

[10] Taylor, J.R. & Taylor Lovell, S., Urban home food gardens in the Global North: Research traditions and future directions. *Agriculture and Human Values*, **31(2)**, pp. 285–305, 2014. https://doi.org/10.1007/s10460-013-9475-1

[11] Rosen, C.J., Bierman, P.M. & Eliason, R.D., *Soil Test Interpretations and Fertilizer Management for Lawns, Turf, Gardens, and Landscape Plants*. University of Minnesota Extension, St. Paul, MN, 2008.

[12] Dewaelheyns, V., Elsen, A., Vandendriessche, H. & Gulinck, H., Garden management and soil fertility in Flemish domestic gardens. *Landscape and Urban Planning*, **116**, pp. 25–35, 2013. https://doi.org/10.1016/j.landurbplan.2013.03.010

[13] Metson, G.S. & Bennett, E.M., Phosphorus cycling in Montreal's food and urban agriculture systems. *PLoS ONE*, **10(3)**, p. e0120726, 2015. https://doi.org/10.1371/journal.pone.0120726

[14] Abdulkadir, A., Leffelaar, P.A., Agbenin, J.O. & Giller K.E., Nutrient flows and balances in urban and peri-urban agroecosystems of Kano, Nigeria. *Nutrient Cycling in Agroecosystems*, **95(2)**, pp. 231–254, 2015. https://doi.org/10.1007/s10705-013-9560-2

[15] Murphy, J. & Riley, J.P., A modified single solution method for the determination of phosphate in natural waters. *Analytical Chimica Acta*, **27**, pp. 31–36, 1962. https://doi.org/10.1016/s0003-2670(00)88444-5

[16] Tilman, D., Global environmental impacts of agricultural expansion: The need for sustainable and efficient practices. *Proceedings of the National Academy of Sciences of the United States of America*, **96(11)**, pp. 5995–6000, 1999. https://doi.org/10.1073/pnas.96.11.5995

[17] Small, G.E., Sisombath, B., Reuss, L., Henry, R. & Kay, A.D., Assessing how the ratio of barley mash to wood chips in compost affects rates of microbial processing and subsequent vegetable yield. *Compost Science & Utilization*, **25(4)**, pp. 272–281, 2017. https://doi.org/10.1080/1065657x.2017.1329038

[18] Amlinger, F., Götz, B., Dreher, P., Geszti, J. & Weissteiner, C., Nitrogen in biowaste and yard waste compost: Dynamics of mobilization and availability – A review. *European Journal of Soil Biology*, **39(3)**, pp. 107–116, 2003. https://doi.org/10.1016/s1164-5563(03)00026-8

[19] Gabrielle, B., Da-Silveira, J., Houot, S. & Michelin, J., Field-scale modelling of carbon and nitrogen dynamics in soils amended with urban waste composts. *Agriculture, Ecosystems & Environment*, **110(3–4)**, pp. 289–299, 2005. https://doi.org/10.1016/j.agee.2005.04.015

[20] Edmondson, J.L., Davies, Z.G., Gaston, K.J. & Leake, J.R., Urban cultivation in allotments maintains soil qualities adversely affected by conventional agriculture. *Journal of Applied Ecology*, **51(4)**, pp. 880–889, 2014. https://doi.org/10.1111/1365-2664.12254

[21] Booth, M.S., Stark, J.M. & Rastetter, E., Controls on nitrogen cycling in terrestrial ecosystems: a synthetic analysis of literature data. *Ecological Monographs*, **75(2)**, pp. 139–157, 2005. https://doi.org/10.1890/04-0988

[22] Chapin, F.S., Matson, P.A. & Vitousek, P.M., Nutrient cycling. *Principles of Terrestrial Ecosystem Ecology*, pp. 259–296, Springer: New York, 2011.

[23] Palmer, L., Urban agriculture growth in US cities. *Nature Sustainability*, **1(1)**, pp. 5–7, 2018. https://doi.org/10.1038/s41893-017-0014-8

[24] Lee-Smith, D., Cities feeding people: An update on urban agriculture in equatorial Africa. *Environment and Urbanization*, **22(2)**, pp. 483–499, 2010. https://doi.org/10.1177/0956247810377383

SUSTAINABLE AGRICULTURE IN ORGANIC WHEAT (*TRITICUM AESTIVUM* L.) GROWING IN ARID REGION

N.S. AL-GHUMAIZ
Department of Plant Production and Protection, College of Agriculture and Veterinary Medicine, Qassim University, Kingdom of Saudi Arabia.

ABSTRACT

Qassim region is considered an urban, agricultural area in the Kingdom of Saudi Arabia. This region is characterized by dominant arid climate. Low fertility soil is considered a major challenge for the sustainable cultivation of wheat (*Triticum aestivum* L.). The objective of this study was to assess some agronomic characteristics of the wheat genotypes grown organically on low fertility soil. The experiments were conducted during the 2010 and 2011 growing seasons using eight bread wheat genotypes growing under conventional and organic farming systems at two different locations. Measurements of the following parameters were collected: chlorophyll content, flag leaf area (cm^2), and Harvest Index. Findings demonstrate a difference between the conventional and organic farming systems in terms of the parameters under study. Results showed that the greatest chlorophyll content was recorded in IC17 genotype (51.3 SPAD) Genotype Sids 12 had the highest FLA (28.0 and 26.3) under the conventional and organic farming systems, respectively. E-line and YR had the highest harvest index under the organic farming system in both the seasons. For sustainable food production in arid regions using the organic farming system, wheat genotypes YR and E-line could be the most suitable.

Keywords: arid region, conventional system, organic system, sustainable agriculture, wheat genotype.

1 INTRODUCTION

Wheat (*Triticum aestivum* L.) is globally important staple cereal because it constitutes the main source of protein and energy for people in most countries [1].

The use of chemical fertilizers in cereal production has increased worldwide [2] due to availability of inexpensive fertilizers [3]. The continued use of chemical fertilizers causes health and environmental hazards such as ground and surface water pollution by nitrate leaching [4]. So, reducing the amount of nitrogen fertilizers applied to a field without nitrogen deficiency will be the main challenge in field management.

Organic agriculture can be defined as a system that prohibits the use of synthetic fertilizers, chemical pesticides, and genetically modified organisms. As such, crop breeding and agronomic research is increasingly focusing on identifying parameters/traits, which, when adopted, can emulate the conditions and results of organic production.

Harvest Index (HI) can be used as a measure of reproductive efficiency [5]. In wheat, HI is an important parameter that can quantify the yield in terms of the total amount of biomass. Another such important parameter for wheat species is the flag leaf area. It has been found that flag leaf photosynthesis in wheat contributes to ~30–50% of the assimilates during grain filling [6]. Maintaining chlorophyll content, however, is the major challenge of organic farming systems that utilize environment-friendly farming practices, to ensure high yield and excellent quality [7].

Several field trials have compared organic and conventional farming systems, and shown that there is a significant lower yield in the organic systems [8], [9]. In a long-term study in Sweden [10] comparing organic and conventional agriculture involving barley (*Hordeum vulgare*), wheat, and some other crops, it was shown that there is a disparity in productivity among the crops, as well as between the organic and conventional farming systems.

DOI: 10.2495/DNE-V14-N1-1-6

Sustainable agriculture in urban areas with poor soil nutrients is a challenge. However, the promising results of our previous study, wherein we identified spring wheat genotype(s) suitable for cultivation under organic farming systems in arid regions [11], encouraged us to extend our study further to investigate the effects of selecting other agronomic characters including harvest Index, chlorophyll content (%) and flag leaf area (cm^2) on wheat productivity.

2 MATERIALS AND METHODS

Organic and conventional farming trials were conducted during the 2010 and 2011 growing seasons at two different locations (Farming systems). The conventional trial was established at Qassim University Agricultural Research Station, and involved the application of standard agricultural practices, including the use of chemical fertilizers as recommended by soil test results. The organic trial was established at the Research Center for Organic Agriculture, which is a certified organic farm near to the conventional trial (26°18′N, 43°46′E, elevation of 648 m). The organic trial was subjected to treatments adopted according to the organic system regulations set by the Ministry of Environment, Water and Agriculture of the Kingdom of Saudi Arabia (KSA).

Results of the analysis of soil chemical and physical properties of the two experimental stations are listed in Table 1. The maximum and minimum air temperatures during the 2010 and 2011 growing seasons (Jan to April) compared to the 27-year average are presented in Figs. 1 and 2, respectively.

Table 1: Results of analysis of the soil chemical and physical properties at the two experimental sites.

	Chemical analyses						Physical properties		
Farming system	K (ppm)	P (ppm)	N (ppm)	OM%	pH	Ec ds/m)	Clay %	Silt %	Sand %
Conventional	34	33.1	15.7	0.14	8.1	5.3	0.4	4.2	94.9
Organic	36.5	22.1	52.5	1.4	7.9	1.9	0.9	4.5	94.5

OM: Organic matter, Ec: Electrical conductivity.

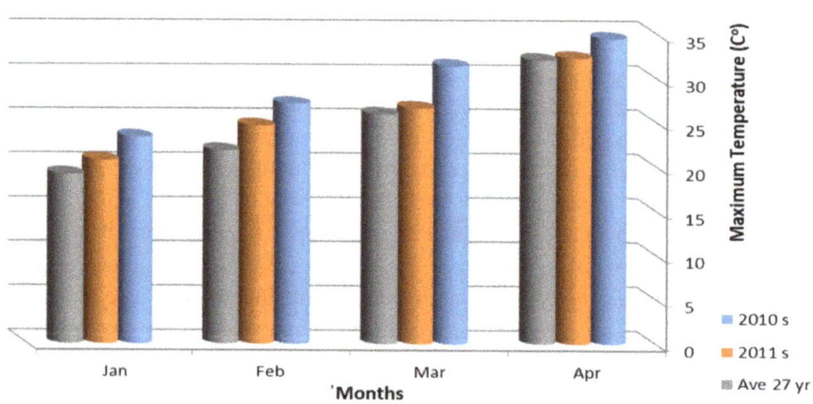

Figure 1: Monthly maximum air temperature during the two growing seasons compared to the 27-year average.

Source: *The General Authority of Meteorology & Environmental Protection-Saudi Arabia. http://www.pme.gov.sa*

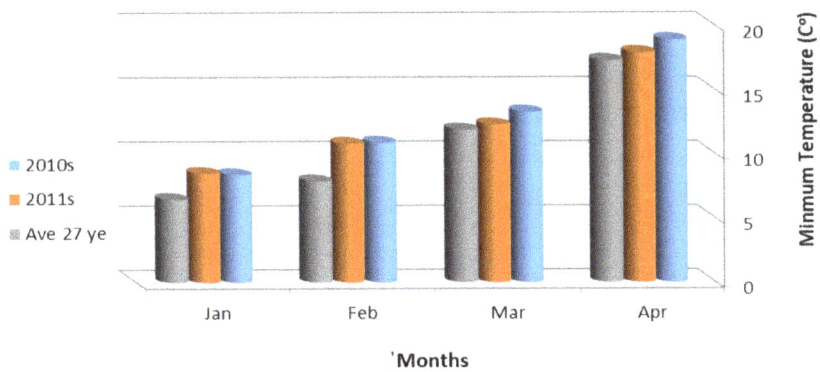

Figure 2: Monthly minimum air temperature during the two growing seasons compared to the 27- year average.

Source: The General Authority of Meteorology & Environmental Protection-Saudi Arabia. http://www.pme.gov.sa

Eight bread wheat genotypes were used: Yocora Rojo (YR; the commercial genotype commonly grown in Saudi Arabia), Sama (the local genotype), the Australian genotypes P3 (AUS-030851) and P5 (AUS-030852), IC8 (Line-2-ICARDA-1st RDRN0607), IC17 (Line-56 ICARDA-1st RDRN0607), and the Egyptian genotypes Early line (E.Line) and Sids 12.

The experiment was set up as a randomized complete block design (RCBD) with four replications using 3-m^2 plots (1.5 m × 2 m). Seeding rates were 45 seeds per plot. The following parameters were measured: chlorophyll content (CHC), flag leaf area (FLA), and HI. After heading, 10 flag leaves were collected from each experimental unit and measured using a LI-3100 (LI-COR, Lincoln, Nebraska, USA) leaf area meter. Chlorophyll pigment was measured at the late booting stage on the top leaf of 10 plants in each experimental unit using a SPAD-501 Chlorophyll Meter (Konica Minolta, Co. Ltd., Japan).

Analysis of variance (ANOVA) was conducted on data of all the measured parameters, and the mean values for the eight wheat genotypes were compared using the least significant differences (LSD) test, with the level of significance set at $P < 0.05$. Coefficient of variation (CV) was listed to measure the precision of the trial. Each location (Farming system) was analyzed as an independent experiment. The collected data for all variables were statistically analyzed using the MSTATC microcomputer program [12].

3 RESULTS AND DISCUSSION

There was a significant interaction between location (Farming system) and year in terms of HI and FLA. Similar interactions were also observed between location and genotype (Table 2). We attribute these interactions primarily to weather conditions varying between locations and between years (Figs 1 and 2), as well as to variations among the genotypes studied. No significant interaction was found between location and genotype in terms of chlorophyll content (Table 2).

3.1 Chlorophyll content

Chlorophyll content was affected by the farming system and genotype individually, although the effect of the interaction of these two factors was not significant (Table 2). Under the con-

Table 2: Effect of farming systems and genotypes on some agronomic traits.

	Chlorophyll content (SPAD)	Flag leaf area (cm^2)	Harvest Index
Treatment	Location (Farming system)		
C-2010	51.9a	10.3c	0.33c
Or-2010	49.8ab	9.8c	0.37b
C- 2011	49.5b	20.8a	0.40a
Or-2011	49.4b	20.8a	0.37b
Sig	**	**	**
Genotype			
Y R	49.5b	14.0c	0.45a
Local	45.8d	18.8a	0.28d
P3	46.9cd	14.2c	0.37c
P 5	47.1cd	15.7b	0.36c
IC 8	47.3cd	10.6e	0.38bc
IC 17	51.3a	16.1b	0.38bc
E. Line	48.1bc	12.1d	0.45a
Sids 12	48.9b	18.9a	0.40b
Sig	**	**	**
L×G	Ns	**	**

C-2010: Conventional 2010, C-2011: Conventional 2011, Or 2010: Organic 2010, Or 2011: Organic 2011.
ns and** = not significant at 0.05 and significant at 0.01 level of probability, respectively
Means in column within each factor designated by the same latter are not statistically different at *0.05* level of probability using the (LSD).

ventional system, the chlorophyll content was the highest during 2010 season (51.9 SPAD). Genotype IC17 demonstrated the greatest chlorophyll content (51.3 SPAD). YR and the Egyptian genotypes (E-line and Sids 12) had higher chlorophyll content comparable to those in the local, and Australian genotypes (P3 and P5).

3.2 Flag leaf Area (cm^2)

The area of flag leaf was not affected by the farming system. Only one genotype (P3) had significantly different FLA when grown under the two farming systems during the 2010 growing season. Genotype Sids 12 had the highest FLA (28.0 and 26.3 cm^2) under the conventional and organic farming systems, respectively, in 2011. FLA varied significantly between the two growing seasons ($P < 0.05$), with the 2011 season recording the highest FLA in both the farming systems. (Table 3).

3.3 Harvest index

There was no significant difference between the conventional and organic farming systems, in terms of HI, with the exception of IC17, which showed a highly significant difference

Table 3: Effect of interaction between the two farming systems and genotypes on flag leaf area (cm^2).

	Flag leaf area (cm^2)			
Genotype	C-2010	Or-2010	C- 2011	Or-2011
Y R	10.9	8.7	19.1	18.3
Local	12.9	13.4	25.4	27.5
P3	10.7	7.0	21.0	21.1
P 5	10.3	10.1	22.1	23.4
IC 8	7.1	8.5	13.0	15.2
IC 17	10.3	12.8	21.6	20.2
E. Line	8.8	7.0	16.3	14.4
Sids 12	11.1	11.1	28.0	26.3
LSD	3.4			

C: conventional, Or: Organic.

Table 4: Effect of interaction between the two farming systems and genotypes on HI.

	Harvest Index			
Genotype	C-2010	Or-2010	C- 2011	Or-2011
Y R	0.41	0.46	0.46	0.45
Local	0.24	0.26	0.28	0.24
P3	0.31	0.34	0.39	0.35
P5	0.30	0.34	0.37	0.32
IC 8	0.30	0.34	0.41	0.37
IC 17	0.30	0.40	0.39	0.34
E. Line	0.39	0.43	0.47	0.48
Sids 12	0.36	0.38	0.41	0.39
LSD	0.05			

C: conventional, Or: Organic.

($P < 0.05$) between the two systems. Interestingly, E-line and YR had the highest HI under the organic farming system in both the seasons. Growing such genotypes can help for sustainable food production in urban regions.

There was a significant difference ($P < 0.05$) between the 2010 and 2011 growing seasons in terms of the HI only under the conventional farming system (Table 4).

4 CONCLUSIONS

Organic farming systems can be adopted for ensuring a clean environment and sustainable crop production. In wheat cultivation, selection of some specific parameters such as chlorophyll content and HI, can be important to ensure the success of organic farming. For sustainable production in arid regions using the organic farming system, wheat genotypes YR and E-line could be the most suitable.

ACKNOWLEDGEMENTS

The author acknowledges the Deanship of Scientific Research at Qassim University for funding this project. I gratefully thanks Dr. Mohammed Motawei for reviewing this paper and Dr. Abdel-Salam Menshawy for his assistance with the statistical analyses. I would like to thank Editage (www.editage.com) for English language editing.

REFERENCES

[1] Bos, C., Juillet, B., Fouillet, H., Turlan, L., Dare, S., Luengo, C., N`tounda, R., Benamouzig, R., Gausseres, N., Tome, D. & Gaudichon, C., Postprandial metabolic utilization of wheat protein in humans. *The American Journal of Clinical Nutrition*, **81(1)**, pp. 87–94, 2005. https://doi.org/10.1093/ajcn/81.1.87

[2] Abril, A., Baleani, D., Casado-Murillo, N. & Noe, L., Effect of wheat crop fertilization on nitrogen dynamics and balance in the Humid Pampas, Argentina. *Agriculture, Ecosystems & Environment*, **119(1–2)**, pp. 171–176, 2007. https://doi.org/10.1016/j.agee.2006.07.005

[3] Graham, P.H. & Vance, C.P., Nitrogen fixation in perspective: an overview of research and extension needs. *Field Crops Research* **65(2–3)**, pp. 93–106, 2000. https://doi.org/10.1016/s0378-4290(99)00080-5

[4] Pimentel, D., Green revolution and chemical hazards. *Science of the Total Environment*, **188(supplement)**, pp. S86–S98, 1996. https://doi.org/10.1016/0048-9697(96)05280-1

[5] Murray, U., Jeff, B. & Matthew, F., Variability in harvest index of grain crops and potential significance for carbon accounting: Examples from Australian agriculture. *Advances in Agronomy*, ed. D.L. Sparks, pp 173–219, Elsevier, 2010.

[6] Sylvester, B.R., Stokes, D.T. & Scott, R.K., A physiological analysis of the diminishing responses of winter wheat to applied nitrogen, 1. Theory. *Aspects of Applied Biology*, pp. 277–287, 1990.

[7] Tilman, D., Cassman, K.G., Matson, P.A., Naylor, R. & Polasky, S., Agriculture sustainability and intensive production practices. *Nature*, **418(6898)**, pp. 671–677, 2002. https://doi.org/10.1038/nature01014

[8] Ryan, M.H., Derrick, J.W. & Dann, P.R., Grain mineral concentrations and yield of wheat grown under organic and conventional management. *Journal of the Science of Food and Agriculture*, **84(3)**, pp. 207–216, 2004. https://doi.org/10.1002/jsfa.1634

[9] Stanhill, G., The comparative productivity of organic agriculture. *Agriculture. Ecosystems & Environment*, **30(1–2)**, pp. 1–26, 1990. https://doi.org/10.1016/0167-8809(90)90179-h

[10] Kirchmann, H., Bergström, L., Kätterer, T. & Gesslein, S., Comparison of long-term organic and conventional crop–livestock systems on a previously nutrient-depleted soil in Sweden. *Agronomy Journal*, **99(4)**, pp. 960–972, 2007. https://doi.org/10.2134/agronj2006.0061

[11] Al-Ghumaiz, N.S., Yield performance quality of eight wheat genotypes under organic and conventional farming systems in Saudi Arabia. *Agriculture and Food*, **2**, pp. 20–24, 2014. [Online] Available: http://www.scientific-publications.net/en/article/1000002/

[12] MSTATC. A microcomputer program for the design. Management, and Analysis of Agronomic Research Experiments. Michigan State University East Lansing, MI, USA. 1990.

HEAT TRANSFER MECHANISMS IN VERTICAL GREEN SYSTEMS AND ENERGY BALANCE EQUATIONS

F. CONVERTINO, G. VOX & E. SCHETTINI
Department of Agricultural and Environmental Science – DISAAT – University of Bari, Bari, Italy.

ABSTRACT

The use of vegetated vertical systems is a sustainable technology for improving the energy efficiency of buildings in cities in order to reduce the energy consumption for air conditioning in summer and to increase the thermal insulation in winter. Increasing urban green infrastructure (UGI) in a city can contribute to improve urban climate in summer reducing buildings surface temperature and urban air especially in Southern Europe. The application of vertical green systems requires the knowledge of the energy performance of the applied greenery system. The choice of the green facades depends on the local climate, water availability, building shape.

The presence of green facades affects the building microclimate all day, by reducing heat waves during the warm periods and heat losses from the building in the cold period.

The heat and mass transfer between the external environment, the green facades and the building surface determine the building microclimate. Solar radiation, long wave infrared radiation, convective heat transfer and evapotranspiration are the main mechanisms of heat transfer in a green façade.

The paper describes the main parameters concerning heat flow in green facades that can be used in simulation models for predicting temperatures in buildings using the external weather conditions as model inputs. The input parameters are: external air temperature and relative humidity, solar radiation, wind velocity and direction, plants and building characteristics.

The green facade was described by a schematic representation, four layers were defined: the green layer, the external surface of the building wall, the internal surface of the building wall, the air inside the building.

The energy balance was defined for each layer and all the terms involved in the energy exchange between the layers were defined as a function of the plant, the weather conditions and the constructive characteristics of the wall.

Keywords: convection, evapotranspiration, green facades, green walls, modelling, solar radiation, urban agriculture.

1 INTRODUCTION

Vertical greening systems are characterized by plants grown on buildings vertical surfaces. This kind of envelope technological solution has a very ancient history, dating back to Babylonians. Over time, the forms of vertical green systems have spread and have changed considerably. Literature is rich in an incredible variety of definitions of all the different types of vertical greening systems [1], [2]. The main difference is between green facades and green walls. Green facades are usually realized with climber plants covering the wall and rooted into the soil or in planter boxes. They can be distinguished into direct green facades, where plants are attached directly to the wall, and indirect green facades, characterized by the presence of a supporting system, such as continuous guides or modular trellises, to assist the upward growth of climbing plants, in this way creating a sort of second skin detached from the wall. Green walls are an alternative and most recent solution to vegetate the vertical walls. They are usually composed of a supporting element, a growing medium, plants and an irrigation system. Supporting elements can be lightweight and permeable screens in which plants are inserted individually (continuous living wall systems) or pre-cultivated modular panels (modular living wall systems) that already contain the growing medium and the irrigation system; they can be fixed directly to the wall or to a supporting frame. The panels or

DOI: 10.2495/DNE-V14-N1-7-18

geotextile felts support plants development. Green walls rapidly create a uniform growth of a wider variety of plant species along large wall surfaces in high buildings [3], [4].

Nowadays, the spread of vertical greening systems is rising. This is due to the growing need of environmental sustainability. In fact, the use of green layers in buildings is a smart way to mitigate several environmental problems and to improve the quality of urban contexts. The advantages deriving from their use and diffusion are many and very remarkable, under various points of view and at different scales. Vertical green layers applied to building walls allow to reduce energy consumption for air conditioning, in summer, and to improve thermal insulation, in winter [3], [5]–[10]. Thus, green vertical systems can be used as a passive technology for enhancing the energy efficiency of buildings and for mitigating the frequency and magnitude of the heat events due to urban heat island (UHI). In comparison to the green roofs, the green vertical systems are the most suitable systems for greening dense urban areas because buildings with a high wall to roof ratio offer large surface areas available for retrofitting [12]. They reduce the ambient temperatures and improve human thermal comfort [4], [10], [11]. Green vertical systems function as a solar shading barrier reducing the heating of the external surface of the building during daytime and function as thermal barrier at nighttime [8], [13]. An important and additional benefit of this living screen, respect to no living ones, arises in relation to solar radiation. In fact, plants produce a cooling effect by intercepting the incident solar radiation; actually, plants partially reflect and partially use solar radiation for the biological activities such as evapotranspiration, photosynthesis, and storing it through the lymphatic system [5], [14]. Evapotranspiration leads to cooling the leaf and the temperature of the surroundings and consequently reduces the cooling load of the buildings [12], [14], [15]. The main advantages given by green vertical systems, in summer, are a lower external and internal surface temperature and an improved indoor thermal comfort [4], [5], [10], [11], [16], [17]. While, in winter, they are able to reduce energy demands for heating, by decreasing the wind velocity and by controlling heat losses [3]. In the functioning of vertical greening systems, significative aspects are the wall orientation and the specific characteristics of the greenery system used [13], [18]–[21]. Nevertheless, only few studies deal with the long-term thermal performance of these systems useful to assess suitable plant species and systems options [4], [5], [13]. Another critical aspect is the poor availability of simulation tools, specifically designed to describe and forecast the vertical green systems energy performance. Many authors use effective thermal resistance as a parameter to compare thermal benefit and characteristic of vertical green systems [22]–[24].

The study and the comprehension of the energy balance in vertical greening systems are essential to better understand, to make the most of the potential and to obtain the maximum benefit from their use. When working with living systems, the reaching of this objective is not so easy, because of the numerous and complex variables involved and because plants respond to environmental conditions in a very complicated way [25]. In the most recent literature the researchers' attention to this problem emerges. Some authors, concentrating on the reduction of the solar radiation on the wall behind the plant, focused on the leaf coverage and proposed a sort of shading or bioshading coefficient [26], [27]. However, although this is one of the most significative influence of the green canopy, it is necessary to evaluate all the other complex and determining processes, such as the radiative, the convective and the latent (i.e. the evapotranspiration) heat transfer. It is possible to find, in literature, different modelling approaches for the heat and mass exchanges, interesting, in particular, the vegetated layer: the big leaf, the dual leaf and the multi-layer canopy model, with an increasing complexity. It is also very common to find, in the models, the thermal-electrical analogy. Among the most recent studies' objectives, there is that to realize complete analytical models and to validate them by empirical measurements. Many attempts were made to describe the evapotranspiration process. Larsen *et al.* [25] proposed an energy balance equation for the foliage, in which the latent contribution derives from the well-known Penmann–Monteith (PM) equation, advised by FAO [28]. The authors considered a glazed façade, to point out the reduction in the direct solar radiation through the green wall. Malys *et al.* [29] proposed a hydrothermal model of vegetated walls, for the implementation in an urban microclimate simulation software, and compared simulation results with experimental data obtained for three green wall samples. Scarpa *et al.* [30] tried to predict the thermal behaviour of living walls, considering two kinds of living walls, one with grass and closed air cavity and another with vertical garden and open cavity. They used the PM equation and managed to deduce the crop coefficient (which distinguishes the considered crop from the hypothetical reference crop), considering the most correct equivalences with the FAO guide, which includes only agricultural species. Evapotranspiration was particularly analysed also by van de Wouw *et al.* [31]. They derived crop coefficient by comparing the reference crop evapotranspiration value, calculated through PM equation, with empirical data, obtained from an experimental setup, designed to monitor the changes in water contents. From these and other recent studies, a close connection between the solar radiation and the functioning of the vertical green systems clearly emerges. This work allows to point out this influence, as well as that of the other thermal contributions.

Therefore, in a vertical green system all the terms of the energy balance must be determined in the most effective way, even if the presence of the living component makes it not a simple task. Among the balance components, the most characteristic is the evapotranspiration, which deserves a particular attention.

To assess the actual effects of the green vertical systems, mathematical models are needed, also for their implementation in simulation software. This implies the necessity to increase the analytical modelling, still at an initial stage, together with the most common empirical approach.

This paper aims to investigate the heat flow mechanisms involved in green facades. Heat balance equations for the different layers, in which the system was schematized, were defined. The attention was focused, specifically, on the energy balance of the vegetate layer.

2 MATERIALS AND METHODS

The thermal exchange in the green facades occurs by convection, conduction and radiation heat transfer. The latent heat flux also contributes to the energy exchange; latent heat arises from the evapotranspiration that is a typical process of the plants. A schematic representation of the green facade is reported in Fig. 1 with the different heat fluxes existing in the system.

The green facade was described by means of energy balance equations corresponding to the different layers [32], [33]: the green layer (gl), the air gap between the green layer and the building external wall (ag), the external surface of the building wall (ew), the internal surface of the building wall (iw) and the air inside the building (ia). For each of these layers, the heat fluxes were modelled and a heat balance equation was written.

The system was simulated by means of a one-dimensional model, considering only the energy flux normal to the wall, thus neglecting the horizontal fluxes of energy.

2.1 The energy balance equations

The layers exchange energy by radiation in the solar wavelength range between 300 and 3000 nm (E) and in the longwave infrared (LWIR) range >3 μm (R), by convection (CV), conduction (C) and evapotranspiration (Φ); energy is also stored into the layers (S).

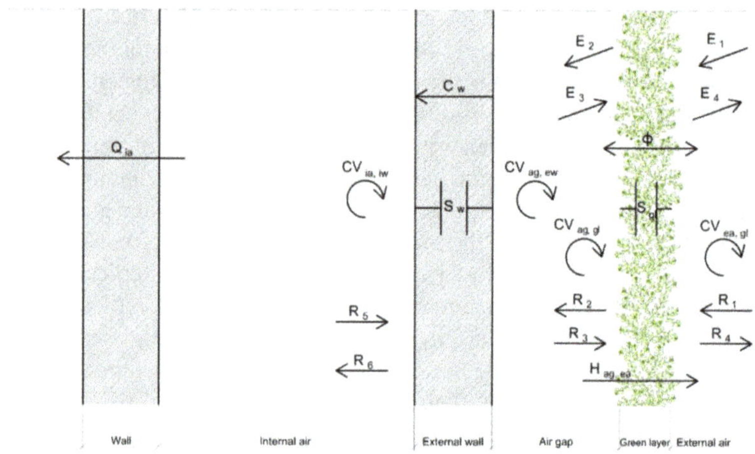

Figure 1: Schematic representation of the green facade and of the energy fluxes for each element.

An energy balance equation was written for each layer (Fig. 1).

The heat balance for the air inside the building is:

$$CV_{ia,iw} - Q_{ia} = 0 \tag{1}$$

The heat balance for the internal surface of the building wall is:

$$\varepsilon_{iw} R_5 - R_6 - CV_{ia,iw} = 0 \tag{2}$$

with:

$$R_5 = \varepsilon_{ir} \sigma T_{ir}^4 \tag{3}$$

$$R_6 = \varepsilon_{iw} \sigma T_{iw}^4 \tag{4}$$

both calculated according to the Stefan–Boltzmann law.

The heat balance for the external surface of the building wall is:

$$E_2 - E_3 + \varepsilon_{ew} R_2 - R_3 - CV_{ag,ew} - C_w - S_w = 0 \tag{5}$$

The terms C_w and S_w express the contributes of heat conduction and storage into the first layer of the building external wall, respectively. Heat storage and conduction in the building wall were considered by using the Fourier's heat equation:

$$\frac{\partial T}{\partial t} = D \frac{\partial^2 T}{\partial z^2} \tag{6}$$

The heat balance for the air of the gap is:

$$CV_{ag,\,ew} - CV_{ag,\,gl} - H_{ag,\,ea} = 0 \tag{7}$$

The term H accounts for the heat exchanged by ventilation between the air in the gap and the external air.

The heat balance for the green layer is described by:

$$E_1 - E_4 - E_2 + E_3 + \varepsilon_{gl}\left(R_1 + R_3\right) - \left(R_4 + R_2\right) + CV_{ea,\,gl} + CV_{ag,\,gl} + S_{gl} - \Phi = 0 \qquad (8)$$

The evapotranspiration term (Φ) is specific of the vegetation.

2.1.1 Energy exchange in the green layer

The attention is, here, focused on each term of the energy balance of eqn (8).

Heat storage

The heat storage in the green layer (S_{gl}) was neglected, since the heat capacity of the vegetated layer is small compared to the other existing fluxes.

Shortwave radiation

E_1 is the solar radiation on a vertical surface, the other E terms can be calculated by means of E_1, the green layer solar reflectivity (ρ_{gl}), transmissivity (τ_{gl}) and absorptivity (a_{gl}) and the external wall surface solar reflectivity (ρ_{ew}):

$$E_2 = \tau_{gl} \cdot E_1 \qquad (9)$$

$$E_3 = \rho_{ew} \cdot \tau_{gl} \cdot E_1 \qquad (10)$$

$$E_4 = \rho_{gl} \cdot E_1 \qquad (11)$$

The algebraic sum of the E terms gives the net solar radiation absorbed by the green layer.

Longwave radiation

The LWIR heat exchanges are:

$$R_1 = R_{sky} + R_{ground} \qquad (12)$$

R_1 includes the flux of LWIR radiation coming from the sky (R_{sky}) and from the ground (R_{ground}) over the solid angle of sky and ground viewed by the green layer, respectively:

$$R_{ground} = \sigma \cdot F_{ground} \cdot T^4_{ground} \qquad (13)$$

$$R_{sky} = \sigma \cdot F_{sky} \cdot T^4_{sky} \qquad (14)$$

R_{ground} and R_{sky} are calculated in accordance with the Stefan-Boltzmann law; with:

$$T_{sky} = 0.0552 \cdot T^{1.5}_{ea} \qquad (15)$$

calculated as suggested by Kindelan [33], in clear sky conditions.

The other R contributes take into account the LWIR emission from the green layer (R_2 and R_4) and from the building external surface (R_3):

$$R_2 = R_4 = \sigma \cdot \varepsilon_{gl} \cdot T^4_{gl} \qquad (16)$$

$$R_3 = \sigma \cdot \varepsilon_{ew} \cdot T^4_{ew} \qquad (17)$$

Convection

The convective heat exchange between the external air and the green layer was calculated according to Papadakis [34], using the Nusselt number for pure forced flow:

$$CV_{ea,gl} = \rho_a \cdot C_{pa} \cdot \left(T_{ea} - T_{gl}\right) \cdot r_e^{-1} \tag{18}$$

The convective heat flux between the air gap and the green layer is:

$$CV_{ag,gl} = h_{ag} \cdot \left(T_{ag} - T_{gl}\right) \tag{19}$$

Evapotranspiration

The latent heat transfer associated to the evapotranspiration process is calculated according to [34], [35]:

$$\Phi = \rho_a \cdot C_{pa} \cdot \left(e_{s,gl} - e_a\right) \cdot \gamma^{-1} \cdot \left(r_s + r_e\right)^{-1} \tag{20}$$

where r_s is the stomatal (or internal) resistance and r_e is the aerodynamic (or external) resistance.

2.2 Application to experimental data

The energy balance model was applied using the data collected in 2015 at the experimental centre of the University of Bari, where a green facade was realized. The wall was made as prototype of a commonly used vertical building closure in Mediterranean civil construction. The wall facing south was built using perforated bricks joined with mortar. *Pandorea jasminoides variegated*, evergreen climbing plant, was chosen as greenery vertical system component. A plant supporting structure made of an iron net was placed at a distance of 15 cm from the vertical wall. The experimental data were collected by means of a meteorological station consisting of a data logger (CR10X, Campbell, Logan, USA) and several sensors for measuring different climatic parameters. The data were measured with a frequency of 60 s, averaged every 15 min and stored in the data logger. The solar radiation normal to the wall was measured using a pyranometer (model 8–48, Eppley Laboratory, Newport, RI, USA) in the wavelength range 0.3–3 μm. The external air temperature was measured by a Hygroclip-S3 sensor (Rotronic, Zurich, Switzerland). The temperature of the external plaster surfaces exposed to the solar radiation was measured using thermistors (Tecno.el s.r.l. Formello, Rome, Italy).

3 RESULTS AND DISCUSSION

The proposed mathematical model aims to simulate the energy behaviour of green facades in different boundary conditions. The model was applied in the case of the green facade realized at the experimental centre of the University of Bari. Two summer clear days characterized by a measured maximum vertical solar radiation equal to 399.2 W/m^2 (17 July 2015) and equal to 608.5 W/m^2 (17 September 2015) were considered.

Figures 2 and 3 show the results of the application of the model to the data recorded on 17 July 2015. Figure 2 shows the main contributes of sensible heat to the energy balance calculated for the green layer: net vertical solar radiation, net longwave radiation and convection; the measured temperature of the vegetation is also shown. The highest contribution to

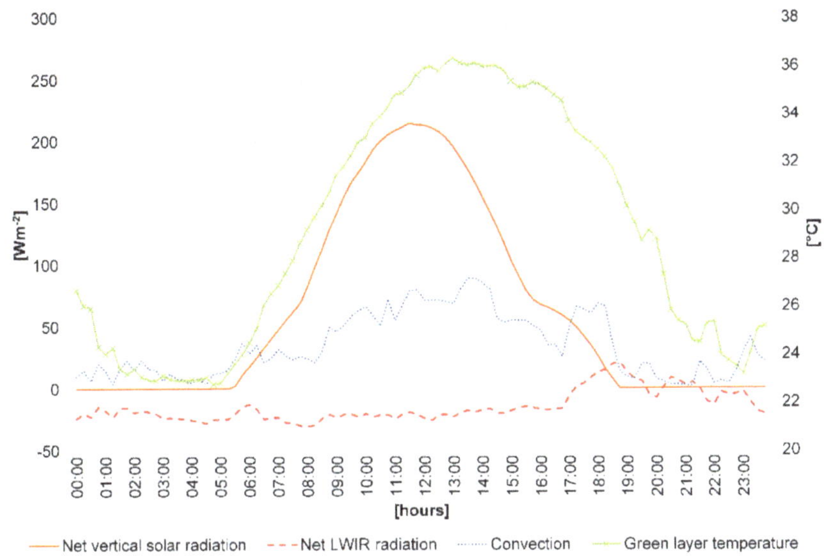

Figure 2: Net vertical solar radiation, net LWIR radiation and convection energy exchange for the green layer, measured green layer temperature (secondary axis), 17 July, 2015.

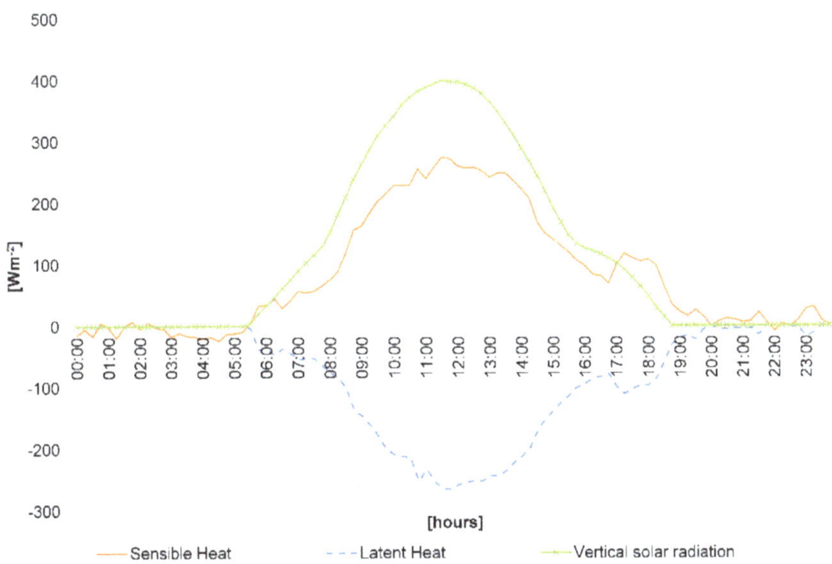

Figure 3: Sensible heat and latent heat exchange for the green layer, measured vertical solar radiation, 17 July, 2015.

the sensible heat absorbed by the vegetation is given by the shortwave radiation. The net LWIR radiation flux has generally negative values, so the green layer loses energy in this way. The convective heat transfer, always positive, has a jagged trend, because it is particularly influenced by the wind speed.

Sensible heat and latent heat for the vegetation are plotted in Fig. 3. The sensible heat is the sum of all the radiative and convective contributes. The latent heat represents the quantity of energy lost by the vegetated layer through the evapotranspiration mechanism. There is a symmetry between the two curves that represent the energy input and output for the layer.

The maximum values of net vertical solar radiation, convection and latent heat, calculated for 17 July 2015, were 214.3 W/m^2, 88.0 W/m^2 and 265.8 W/m^2, respectively.

Similar results were pointed out applying the model to the data recorded on 17 September 2015 (Figs 4 and 5), with the exception of the convective heat transfer that was sometimes negative. The maximum values of net vertical solar radiation, convection and latent heat, calculated for 17 September 2015 were 326.7 W/m^2, 106.5 W/m^2 and 344.7 W/m^2, respectively.

Shortwave radiative energy influences the evapotranspiration and then the latent heat flow (Figs 3 and 5). The solar term seems to be confirmed as the most decisive for latent exchange. Malys *et al.* [29] by comparing latent heat flux values between a sunny and a cloudy day, for green wall samples in Geneva (Switzerland), suggested that solar radiation is the most sensitive parameter in latent heat calculations. Hoelscher *et al.* [36] in the study conducted in Berlin (Germany) on three building facades found that the daily sap flow was influenced by incoming shortwave radiation, air temperature and relative humidity, but that the first was the most influencing factor. Van de Wouw *et al.* [31] analysing two living wall systems in Eindhoven (Netherlands) showed a linear correlation between solar radiation and evapotranspiration and stated that the deviation from linear correlation was due to variations in other climatic factors, i.e. wind speed, air relative humidity and temperature.

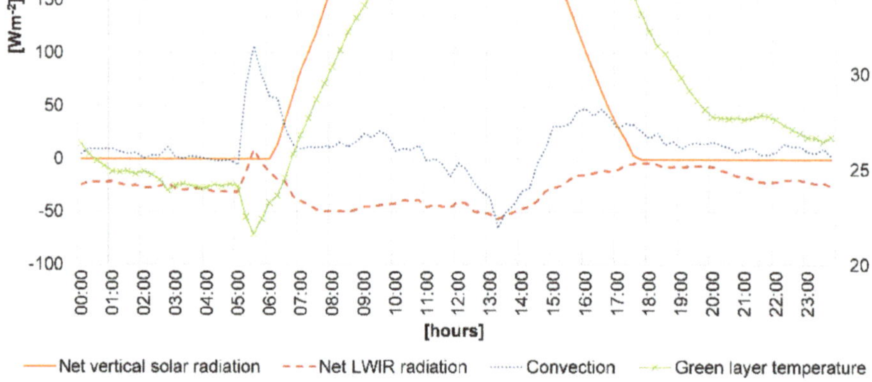

Figure 4: Net vertical solar radiation, net LWIR radiation and convection energy exchange for the green layer, measured green layer temperature (secondary axis), 17 September, 2015.

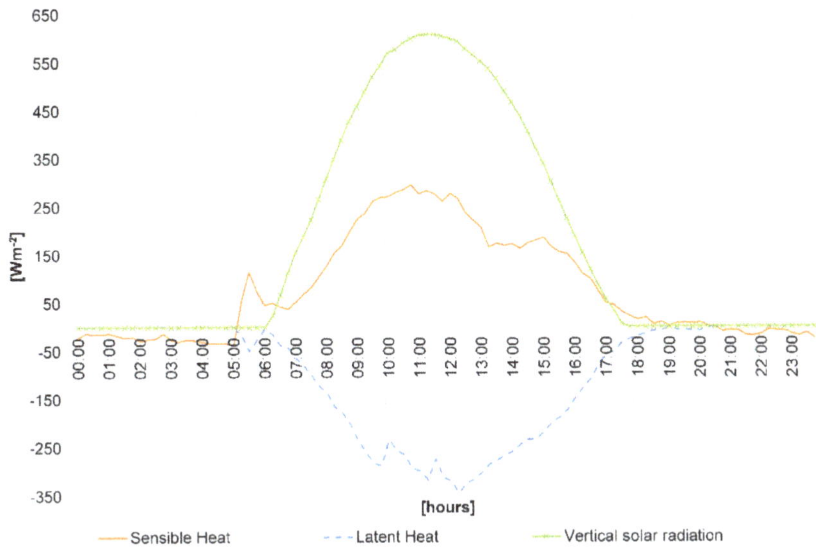

Figure 5: Sensible heat and latent heat exchange for the green layer, measured vertical solar radiation, 17 September, 2015.

NOMENCLATURE

C	heat transfer by conduction $[Wm^{-2}]$	r	resistance $[sm^{-1}]$
C_{pa}	specific heat of air at constant pressure $[Jkg^{-1}K^{-1}]$	S	heat storage $[Wm^{-2}]$
CV	heat transfer by convection $[Wm^{-2}]$	T	temperature [K]
D	thermal diffusivity $[m^2s^{-1}]$	t	time [s]
E	solar radiation $[Wm^{-2}]$	γ	psychrometric constant = 66.7 $[PaK^{-1}]$
e	air vapour pressure [Pa]	ε	emissivity [%]
e_s	air vapour pressure at saturation [Pa]	ρ	solar reflectivity [%]
F	view factor	ρ_a	air density $[kgm^{-3}]$
H	heat transfer by ventilation $[Wm^{-2}]$	σ	Stefan-Boltzmann constant $= 5.6697 \times 10^{-8}$ $[Wm^{-2}K^{-4}]$
h	convective coefficient $[Wm^{-2}K^{-1}]$	τ	solar transmissivity [%]
Q	heat exchanges for air conditioning $[Wm^{-2}]$	Φ	evapotranspiration $[Wm^{-2}]$
R	heat transfer by LWIR radiation $[Wm^{-2}]$		

Subscripts

a	air	ir	internal room
ag	air gap	ew	external surface of the wall
e	aerodynamic or external	iw	internal surface of the wall
ea	external air	s	stomatal or internal
gl	green layer	w	external building wall
ia	internal air		

4 CONCLUSIONS

The study investigated the dynamics of heat transfer in the green facade and the energy balance of the elements that compose it. A heat balance equation was written for each layer.

The proposed energy balance considers all the thermal and physical processes: fluxes of shortwave radiation, longwave radiation, convection, conduction, heat storage and evapotranspiration. The study focuses its attention on the thermal behaviour of the green layer and on its energy balance. A specific objective is to understand the evapotranspiration trend with respect to all the other terms of the balance. The theoretical model was applied to measured data, recorded on an experimental green facade. The analysis of the single terms of the balance pointed out how their contributions influence the overall thermal balance of the green facade. The solar radiation is the most influencing parameter.

Overall, the developed model can be considered able to forecast the thermal behaviour of green facades. The obtained results could be implemented in models, simulating the building microclimate, to define useful tools specifically designed for green facades.

ACKNOWLEDGEMENTS

The contribution to programming and conducting this research must be equally shared between the Authors.

The present work has been carried out under the ‘Piano triennale della Ricerca 2015–2017 nell’ambito del Sistema Elettrico Nazionale, Progetto D.1 ‘Tecnologie per costruire gli edifici del futuro’, Obiettivo: Tecnologie “green” per gli edifici, Piano Annuale di Realizzazione (PAR) 2017’, Accordo di Programma Ministero dello Sviluppo Economico – ENEA funded by the Italian Ministry of Economic Development.

REFERENCES

[1] Medl, A., Stangl, R. & Florineth F., Vertical greening systems – A review on recent technologies and research advancement. *Building and Environment*, **125**, pp. 227–239, 2017. https://doi.org/10.1016/j.buildenv.2017.08.054

[2] Safikhani, T., Abdullah, A.M., Ossen, D.R. & Baharvand, M., A review of energy characteristic of vertical greenery systems. *Renewable and Sustainable Energy Reviews*, **40**, pp. 450–462, 2014. https://doi.org/10.1016/j.rser.2014.07.166

[3] Manso, M. & Castro-Gomes, J., Green wall systems: A review of their characteristics. *Renewable and Sustainable Energy Reviews*, **41**, pp. 863–871, 2015. https://doi.org/10.1016/j.rser.2014.07.203

[4] Pérez, G., Coma, J., Martorell, I. & Cabeza, L.F., Vertical Greenery Systems (VGS) for energy saving in buildings: A review. *Renewable and Sustainable Energy Reviews*, **39**, pp. 139–165, 2014. https://doi.org/10.1016/j.rser.2014.07.055

[5] Bianco, L., Serra, V., Larcher, F. & Perino, M., Thermal behaviour assessment of a novel vertical greenery module system: first results of a long-term monitoring campaign in an outdoor test cell. *Energy Efficiency*, **10(3)**, pp. 625–638, 2016. https://doi.org/10.1007/s12053-016-9473-4

[6] Santamouris, M., Cooling the cities – A review of reflective and green roof mitigation technologies to fight heat island and improve comfort in urban environments. *Solar Energy*, **103**, pp. 682–703, 2012. https://doi.org/10.1016/j.solener.2012.07.003

[7] Schettini, E., Blanco, I., Scarascia Mugnozza, G., Campiotti, C.A. & Vox, G., Contribution of green walls to building microclimate control. *Proceedings of the 2nd International Symposium on Agricultural Engineering (ISAE 2015)*, V-53–V-60, pp. 257–264, 2015.

[8] Vox. G., Blanco, I. & Schettini, E., Green facades to control wall surface temperature in buildings. *Building and Environment*, **129**, pp. 154–166, 2018. https://doi.org/10.1016/j.buildenv.2017.12.002
[9] Schettini, E., Blanco, I., Campiotti, C.A., Bibbiani, C., Fantozzi, F. & Vox, G., Green control of microclimate in buildings. *Agriculture and Agricultural Science Procedia*, **8**, pp. 576–582, 2016. https://doi.org/10.1016/j.aaspro.2016.02.078
[10] Cameron, R.W.F., Taylor, J.E. & Emmett, M.R., What's 'cool' in the world of green façades? How plant choice influences the cooling properties of green walls. *Building and Environment*, **73**, pp. 198–207, 2014. https://doi.org/10.1016/j.buildenv.2013.12.005
[11] Norton, B.A., Coutts, A.M., Livesley, S.J., Harris, R.J., Hunter, A.M. & Williams, N.S.G., Planning for cooler cities: A framework to prioritise green infrastructure to mitigate high temperatures in urban landscapes. *Landscape and Urban Planning*, **134**, pp. 127–138, 2015. https://doi.org/10.1016/j.landurbplan.2014.10.018
[12] Raji, B., Tenpierik, M.J. & Van Den Dobbelsteen, A., The impact of greening systems on building energy performance: A literature review. *Renewable and Sustainable Energy Reviews*, **45**, pp. 610–623, 2015. https://doi.org/10.1016/j.rser.2015.02.011
[13] Jim, C.Y. & He, H., Estimating heat flux transmission of vertical greenery ecosystem. *Ecological Engineering*, **37(8)**, pp. 1112–1122, 2011. https://doi.org/10.1016/j.ecoleng.2011.02.005
[14] Bowler, D.E., Buyung-Ali, L., Knight, T.M. & Pullin, A.S., Urban greening to cool towns and cities: A systematic review of the empirical evidence. *Landscape and Urban Planning*, **97(3)**, pp. 147–155, 2010. https://doi.org/10.1016/j.landurbplan.2010.05.006
[15] Wong, N.H., Kwang Tan, A.Y., Chen, Y., Sekar, K., Tan, P.Y., Chan, D., Chiang, K. & Wong, N.C., Thermal evaluation of vertical greenery systems for building walls. *Building and Environment*, **45(3)**, pp. 663–672, 2010. https://doi.org/10.1016/j.buildenv.2009.08.005
[16] Coma, J., Pérez, G., de Gracia, A., Burés, S., Urrestarazu, M. & Cabeza, L.F., Vertical greenery systems for energy savings in buildings: A comparative study between green walls and green facades. *Building and Environment*, **111**, pp. 228–237, 2017. https://doi.org/10.1016/j.buildenv.2016.11.014
[17] Campiotti, C.A., Schettini, E., Alonzo, G., Viola, C., Bibbiani, C., Scarascia Mugnozza, G., Blanco, I. & Vox, G., Building green covering for a sustainable use of energy. *Journal of Agricultural Engineering*, **44(2s)**, pp. 253–256, 2013. https://doi.org/10.4081/jae.2013.292
[18] Pérez, G., Rincón, L., Vila, A., González, J.M. & Cabeza, L.F., Green vertical systems for buildings as passive systems for energy savings. *Applied Energy*, **88(12)**, pp. 4854–4859, 2011. https://doi.org/10.1016/j.apenergy.2011.06.032
[19] Perini, K., Ottelé, M., Fraaij, A.L.A., Haas, E.M. & Raiteri, R., Vertical greening systems and the effect on air flow and temperature on the building envelope. *Building and Environment*, **46(11)**, pp. 2287–2294, 2011. https://doi.org/10.1016/j.buildenv.2011.05.009
[20] Vox, G., Blanco, I., Campiotti, C.A., Giagnacovo, G. & Schettini, E., Vertical green systems for buildings climate control. *Proceedings of the 43rd International Symposium - Actual Tasks on Agricultural Engineering*, Sveučilište u Zagrebu, Agronomski fakultet, Zavod za mehanizaciju poljoprivrede. Opatija, Croatia, 24-27/02/2015, pp. 723–732, 2015.
[21] Djedjig, R., El Ganaoui, M., Belarbi, R. & Bennacer, R., Thermal effects of an innovative green wall on building energy performance. *Mechanics & Industry*, **18(1)**, p. 104, 2017. https://doi.org/10.1051/meca/2016015

[22] Vox, G., Blanco, I., Fuina, S., Campiotti, C.A., Scarascia Mugnozza, G. & Schettini, E., Evaluation of wall surface temperatures in green facades. *Proceedings of the Institution of Civil Engineers – Engineering Sustainability*, **170(6)**, pp. 334–344, 2017. https://doi.org/10.1680/jensu.16.00019

[23] Susorova, I., Angulo, M., Bahrami, P. & Stephens, B., A model of vegetated exterior facades for evaluation of wall thermal performance. *Building and Environment*, **67**, pp. 1–13, 2013. https://doi.org/10.1016/j.buildenv.2013.04.027

[24] Kontoleon, K.J. & Eumorfopoulou, E.A., The effect of the orientation and proportion of a plant-covered wall layer on the thermal performance of a building zone. *Building and Environment*, **45(5)**, pp. 1287–1303, 2010. https://doi.org/10.1016/j.buildenv.2009.11.013

[25] Flores Larsen, S., Filippín, C. & Lesino, G., Modeling double skin green facades with traditional thermal simulation software. *Solar Energy*, **121**, pp. 56–67, 2015. https://doi.org/10.1016/j.solener.2015.08.033

[26] Wong, N.H., Tan, A.Y.K., Tan, P.Y. & Wong, N.C., Energy simulation of vertical greenery systems. *Energy and buildings*, **41(12)**, pp. 1401–1408, 2009. https://doi.org/10.1016/j.enbuild.2009.08.010

[27] Ip, K., Lam, M. & Miller, A., Shading performance of a vertical deciduous climbing plant canopy. *Building and Environment*, **45(1)**, pp. 81–88, 2010. https://doi.org/10.1016/j.buildenv.2009.05.003

[28] Allen, R.G., Pereira, L.S., Raes, D. & Smith, M., Crop evapotranspiration – guidelines for computing crop water requirements, FAO, 1998.

[29] Malys, L., Musy, M. & Inard, C., A hydrothermal model to assess the impact of green walls on urban microclimate and building energy consumption. *Building and Environment*, **73**, pp. 187–197, 2014. https://doi.org/10.1016/j.buildenv.2013.12.012

[30] Scarpa, M., Mazzali, U. & Peron, F., Modeling the energy performance of living walls: Validation against field measurements in temperate climate. *Energy and Buildings*, **79**, pp. 155–163, 2014. https://doi.org/10.1016/j.enbuild.2014.04.014

[31] Van de Wouw, P.M.F., Ros, E.J.M. & Brouwers, H.J.H., Precipitation collection and evapotranspiration of living wall systems: A comparative study between a panel system and a planter bow system. *Building and Environment*, **126**, pp. 221–237, 2017. https://doi.org/10.1016/j.buildenv.2017.10.002

[32] Kimball, B.A., Simulation of the energy balance of a greenhouse. *Agricultural Meteorology*, **11**, pp. 243–260, 1973. https://doi.org/10.1016/0002-1571(73)90067-8

[33] Kindelan, M., Dynamic modelling of greenhouse environment. *Transaction of the ASAE*, **23(5)**, pp. 1232–1239, 1980. https://doi.org/10.13031/2013.34752

[34] Papadakis, G., Frangoudakis, A. & Kyritsis, S., Experimental investigation and modelling of heat and mass transfer between a tomato crop and the greenhouse environment. *Journal of Agricultural Engineering Research*, **57(4)**, pp. 217–227, 1994. https://doi.org/10.1006/jaer.1994.1022

[35] Monteith, J.L. & Unsworth, M., *Principles of Environmental Physics*, Edward Arnold: Great Britain, 1973.

[36] Hoelscher, M.T., Nehls, T., Jänicke, B. & Wessolek, G., Quantifying cooling effects of façade greening: Shading, transpiration and insulation. *Energy and Building*, **114**, pp. 283–290, 2016. https://doi.org/10.1016/j.enbuild.2015.06.047

URBAN GREEN INDICATORS: A TOOL TO ESTIMATE THE SUSTAINABILITY OF OUR CITIES

C.M. TUDORIE[1], E. GIELEN[2], M. VALLÉS-PLANELLS[3] & F. GALIANA[3]
[1] Universitat Politècnica de València (UPV) Spain.
[2] School of Civil Engineering. Department of Urbanism. Universitat Politècnica de València, Spain.
[3] School of Agricultural Engineering and Environment. Department of Rural and Agrifood Engineering. Universitat Politècnica de València, Spain.

ABSTRACT

In Europe, more than 70% of the population lives in an urban area. All the challenges related to land use conflicts, demographic changes, climate resilience and human well-being are concentrated inside the cities, since the population is already living in urban areas, which are more and more compact and dependent on grey infrastructure. In this context, urban green infrastructure represents a sustainable solution to maintain the benefits and services provided by urban ecosystems and an efficient urban planning tool to face the urban challenges.

The sustainability of our cities can be reached adopting an innovative vision using the concept of ecosystem services of the urban green infrastructure. Different initiatives to assess the benefits provided by green infrastructure have emerged in the last decade. However, very few take into account the whole range of services provided by urban green infrastructure.

The present article provides a systematic search and synthesis of the most important literature to review indicators of urban green infrastructure. The main goal is to give an insight of how urban green infrastructure is measured in practice. Results show the set of ecosystem services that are being considered when assessing sustainability of green infrastructure and identify the most recurrent indicators at the different scales. This work is expected to contribute to the improvement of the evaluation of green infrastructure effectiveness for providing benefits for urban dwellers.

Keywords: ecosystem services, green space indicators, sustainability, urban challenges, urban green infrastructure.

1 INTRODUCTION

Concepts like sustainability and resilience are the main challenges of the city's cores characterized by continue densification, where daily enjoyment of green urban space is becoming a privilege. Sustainability tries to align social science, environmental preoccupation, civic engineering and future technology. According to United States Environmental Protection Agency [1] sustainability refers to the mechanisms of natural systems, its diversity and capacity of producing everything, which is essential for the ecology balance. It also acknowledges the overexploitation of natural resources by inhabitants in order to achieve a modern lifestyle. Sustainable development includes terms of three pillars, environment, economy and society [2].

On one hand, sustainability is important for the cities to reach the 17 sustainable goals established by United Nations Commission [3]. On the other hand, different conferences, reports, global movements and initiatives concerning the future health of our planet testify to the importance of city sustainability. For instance, Habitat III [4] aimed to warn the society about the harmful and negative economic growth and globalization consequences on the environment and natural resources; Green Economy Initiative (GEI) is a global movement whose purpose is to promote "green governments' economies and to reduce the threats caused by ecological crisis. United Nations Development Programme (UNDP) is fully supporting the implementation of New Urban Agenda by launching Sustainable Urbanization Strategy

DOI: 10.2495/DNE-V14-N1-19-29

[5], through which cities could become sustainable solutions for conflicts, demographic and environmental challenges [1]. Biodiversity loss and ecosystem degradation represent the main cause of the urgency of sustainable solutions and within the sustainable development, green infrastructure (GI) occupied its own niche.

GI is an alternative Nature Based Solution (NBS) to grey infrastructure for the alarming problems faced by cities and citizens and its goal is to achieve the improvement of urban development [6]. According to European Commission [7], GI is strategically planned network of natural and semi-natural areas with other environmental features designed and managed to deliver a wide range of ecosystem services (ES) such as water purification, air quality, space for recreation and climate mitigation and adaptation. When this new concept appeared, as derived from landscape ecology, it included only the natural and semi-natural networks [8], but currently, green infrastructure focal point has moved to cities and became the backbone of urban green infrastructure. Urban GI (UGI) provides environmental, economic and social benefits. Due to its multi-functionality, UGI could be proposed as useful urban design tool and landscape management process integrating many different disciplines and concepts [9].

In addition, UGI represents an important agent against climate change. Greenery is promoted for its potential cooling effect of trees and green spaces. According to Benedict *et al.* [10], the function of green infrastructure is to be a supporting conservation goals tool, to outline the ecological function and its role as a connective ecological element. Besides cultural function, green arteries facilitate habitat for urban fauna and flora, providing in the same time air regulation and seed dispersal.

UGI is a big supporter of sustainable and healthy lifestyles, urban livability and wellbeing [11]. GI should be seen as a facilitator of increased mobility, health and education, economical stimulator, involving leisure and social facilities for different stakeholders.

Nowadays, there is an increasing concern of the relevance of providing evidences of benefits, delivered by green infrastructure, in order to prove that an increase of urban green spaces would provide a proportionally larger number of ecological, economic and social benefits and services. For this purpose, indicators that allow the assessment of the performance of green infrastructure are needed [14]. The main goal of this work is to analyze how urban green infrastructure benefits are practically measured.

The objectives of this article are to identify the ES considered in literature to measure the benefits provided by the UGI; to point out integrated indicators used in literature for the assessment of ES; to discover if GI could be a beneficial territorial planning tool and finally to find out if environmental indicators measure only the environmental benefits or there is any interdependence between the social, economic and ecological aspects.

2 METHODOLOGY

First, we look for UGI indicators in the literature that were applied to indicate the effectiveness of ES generated by green spaces. The urban green infrastructure indicators constitute a collection of data from existing databases, which could be simple recommendations of sustainable indicators made by researchers, agencies, organizations or data collection from cities, looking for the case studies with promising results, after using the indicators to test different theories at city scale.

An extensive literature review of the topic, urban green infrastructure indicators, was carried out by using systematically different search engines like Scopus, Web of Science, Google Scholar and electronic libraries connected to UPV.

To help finding reliable references for the research, the identification of keywords was necessary. The set of keywords identified that helped on research topic proposed are: ecosystem services; green space indicators; sustainability; urban challenges; urban green infrastructure.

The searching process was limited to studies published during a fixed period (2000–2018) and only the academic articles and grey informs were chosen. There has been a significant increase of published literature on this topic in the last 3 years.

Second, indicators have been classified according to Common International Classification of Ecosystem Services (CICES) [12] created by European Environment Agency (EEA) [13]. CICES Classification is divided into three themes: Provisioning, Regulation and Maintenance and Cultural, each of them having nested service classes, groups and types (Table 1). Some indicators could not be included in any of the classes. That is why other inspiring sources were necessary to classify the indicators that could evaluate cultural services [9], [14]. Valles-Planells *et al.* [9] proposed a new classification of landscape services and their related concepts and also added new services. In this article, from this second classification, were chosen three classes (mental health, physical health-associated with active enjoyment and recreation- and social fulfillment grouped with social interactions), in order to range the ambiguous indicators, left without an exact framing. EKLIPSE [14] was the last reference, which helped to range other two classes of cultural indicators, participatory planning and governance (seen as challenge number 7) and accessibility, availability and proximity (seen as challenge 4: green space management). Even some researchers [14–16] proposed a hypothetical cultural class, formed by economic services, almost all indicators which measure the monetary value of economic services, were decided to be attached to provisioning services. They represent a real and quantified benefit, e.g. value of avoided grey infrastructure design (construction and management cost, increase in property values [16] and potential for economic opportunities and green jobs [14]). A new service group, related with water quality called regulation of drought was attached to the

Table 1: Structure of CICES classification [12] (number of classes and groups) and definition of Ecosystem services theme.

Ecosystem services	N. Classes	Groups	Definition
Provisioning	3	8	'All nutritional, non-nutritional material and energetic outputs from living systems as well as abiotic outputs (including water)'. It is make the difference between biotic and abiotic outputs.
Regulation and maintenance	4	11	The division level covers the transformation of biochemical or physical inputs in form of wastes, toxic substances and other nuisances and the regulation of physical, chemical, biological conditions, which are benefits, less tangibles than goods but important for human beings.
Cultural	2	4	Includes 'non-material and non-consumptive outputs of ecosystems (biotic and abiotic), that affect physical and mental states of people'.

new classification of ES in order, to classify the indicators proposed by Schyns *et al.* [17], e.g. agricultural drought indicators, agricultural suitability under rain-fed condition, aridity indicators.

After the reclassification, service classes count 18, instead of 9, the initial number of service classes proposed by CICES [12].

In the following, the proposed urban green infrastructure indicators will be analyzed. First, it has been presented a table with a new combination of ecosystem services ranged gathering the two existent classifications mentioned before and new classes and groups inspired from different references. Indicators are classified in function of ecosystem services diversity regarding nested categories and subcategories. Then, the repeated indicators found in the literature will be described in different tables and figures, which indicators use quantitative and qualitative data and their proportion in the proposed classification and finally, examples of integrated indexes will be mentioned.

3 RESULTS AND DISCUSSION

From all the studies found, UGI indicators were subject of 21 references [13–32]. They include qualitative and quantitative measurements of all the categories of ES and contain 175 ecosystem service indicators, able to measure the benefits provided by UGI. Generally, the idea of green indicators is seen as a possible solution to quantify urban challenges or a hypothetical urbanism planning tool. Therefore, the majority of references are making proposals, without having real results of measured benefits provided by urban green infrastructure, e.g. EKLIPSE [14], Pakzad and Osmond [16], Valles *et al.* [9], EPA [1], Artmann *et al.* [31] and McKinsey *et al.* [33]. Nevertheless, there are also articles, which report their results, e.g. Refs. [13], [16], [19–21], [26], [27], [31], [32].

UGI indicators have been ranged using two existing classifications [9], [12], where some new classes and groups have been added (Table 2). The first column of Table 2 contains all three main categories of ecosystem services (marked with letters from A to C). The second level, class service is numbered with Arabic numerals and the last one, which represents service group, is not marked or numbered. The second column reveals the number of indicators found for each class or group and, in the last one, author's name is indicated.

From the total number of indicators, 24 indicators (13.71%) could assess the effectiveness of provisioning services, 83 indicators (38.86%) are used to measure regulating and maintenance services and 68 (47.43%) indicators could estimate cultural services (Fig. 1). The most repeated indicators are shown in Table 3.

From 175 indicators, 20 indicators considered for all three categories of ES have been repeated several times. For instance, carbon storage, carbon sequestration, biodiversity, structural connectivity, total green area or proximity to green spaces, have been mentioned or applied several times in literature [15], [18–24], [25], [27], [32]. With regard to regulation and maintenance services, the most repeated indicators (14.46%) are related to atmospheric regulation, water quality, diversity and functional connectivity. Within cultural services, the most frequent indicators are connected to accessibility, availability, proximity of green spaces. Finally, only one indicator is repeated within the group of provisioning services, which is related to monetary value of properties.

Concerning the data type (Fig. 2), the percentage of quantitative indicators to estimate the benefits of provisioning (11.43%) and regulation and maintenance services (40%) are higher than qualitative indicators percentage (2.29% and 7.43%). When it comes to cultural indicators, qualitative data dominates (21.14%).

Table 2: Urban Ecosystem services (UES) and number of associated indicators (NAI) based on CICES [12], Valles-Planells [9] and EKLIPSE [14].

Urban ecosystem services	NAI	References
PROVISIONING		
A.1. Nutrition		7 authors/ 24 indicators
A.1.1 Terrestrial plan and animal foodstuffs	6	Artmann *et al.* 2017; EKLIPSE, 2016; MAES, 2016; Pakzad *et al.* 2016; EPA, 2012; Fischer *et al.* 2007; Rueda, 2007
A.1.2.Freshwater plant and animal foodstuff	3	
A.1.3 Potable water	3	
A.2. Materials	1	
A.3. Energy	3	
A.4. Monetary values in property/jobs/NBS	8	
REGULATING AND MAINTENANCE		12 authors/83 indicators
B.1. Regulation of waste	10	Baro, 2016; EKLIPSE, 2016; Leff, 2016; MAES, 2016; Pakzad *et al.* 2016; Schyns *et al.* 2015; Barrico *et al.* 2012; EPA, 2012; Agencia d'Ecología Urbana de Barcelona, 2010; Fischer *et al.* 2007; Rueda, 2007; Whitford *et al.* 2001
B.2. Flow regulation (air, water)	18	
B.3. Regulation of physical environment	19	
B.3.1.Atmospheric regulation		
B.3.2.Water quality (regulation of drought)	19	
B.3.3.Pedogenesis and soil quality regulation	12	
B.4. Regulation of biotic environment	5	
B.4.1.Lifecycle maintenance & habitat protection	18	
B.4.2.Gene pool protection	1	
CULTURAL		12 authors/68 indicators
C.1. Symbolic		Artman *et al.* 2017; Fischer *et al.* 2017; Grunewald *et al.* 2017; EKLIPSE, 2016; Leff, 2016; MAES, 2016; Pakzad *et al.* 2016; Ioja *et al.* 2014; EPA, 2012; Agencia d'Ecología Urbana de Barcelona , 2010; Rueda, 2007; Whitford *et al.* 2001
C.1.1.Aesthetic heritage	7	
C.2. Accessibility, availability, proximity	20	
C.3. Recreation and community activities		
C.3.1.Charismatic or iconic wildlife or habitats	10	
C.4. Information and knowledge		
C.4.1.Educational-subject matter for wildlife programmes & books etc.	3	
C.5. Mental health	2	
C.6. Physical health/active enjoyment/recreation	8	
C.7. Social fulfillment /social interactions	11	
C.8. Participatory planning and governance impacts	7	

The biggest number of indicators accounted (70) are quantitative indicators and were proposed to assess regulation and maintenance services as they measure mostly variables of physical environment (e.g. carbon sequestration and storage, greenhouse emissions, run-off coefficient, urban heat island effect, wind speed, Simpson's diversity index or population index). Cultural services category is very diverse involving indicators from multiple domains easier to estimate with qualitative data than quantitative measurements, that is why qualitative indicators predominate. Some examples are social values for urban ecosystems and

Table 3: Repeated indicators used to assess Ecosystem Services (ES), number of times being repeated in literature (T) and the associated references.

ES	Repeated indicators	T	References
A.4.	Increased property values	2	Pakzad and Osmond, 2016; EKLIPSE, 2016
B.2.	Run-off coefficient	2	Whitford *et al.* 2001, EKLIPSE, 2016
B.3.1.	Carbon storage Carbon sequestration Air quality (pollutant removal) Share of emissions captured/ sequestered by vegetation	7	EKLIPSE, 2016; Pakzad and Osmond, 2016; Baro , 2016; Whitford *et al.* 2001;Agencia d'Ecología Urbana de Barcelona , 2010
	Temperature moderation	3	
	Value of air pollutant removal/avoidance	2	EKLIPSE, 2016; Pakzad and Osmond, 2016
	Monetary values: value of air pollution reduction and carbon sequestration	2	Pakzad and Osmond, 2016; EKLIPSE, 2016
B.3.2.	Water consumption	2	Schyns *et al.* 2015; Agencia d'Ecología Urbana de Barcelona , 2010
B.4.	Diversity, biodiversity (species richness, abundance)	3	Barrico *et al.* 2012; Gregory *et al.* 2004; Krasny *et al.* 2016;Agencia d'Ecología Urbana de Barcelona ,2010; EKLIPSE, 2016; Cohen *et al.* 2012; Leff, 2016; Fischer *et al.* 2017;
B.4.1.	Functional connectivity	3	EKLIPSE, 2016;Agencia d'Ecología Urbana de Barcelona , 2010; Rueda, 2007
	Permeability index (soil biotic index)	2	Agencia d'Ecología Urbana de Barcelona, 2010; Rueda, 2007
C.1.1.	Spatial perception of urban green	2	Agencia d'Ecología Urbana de Barcelona, 2010; EKLIPSE, 2016
C.2.	Green-space accessibility	3	Grunewald *et al.* 2017; EKLIPSE, 2016; Rueda, 2007
	Total area of green space	5	Rueda, 2007; Whitford *et al.* 2001; Istat, 2013; Grunewald *et al.* 2017; Barrico *et al.* 2012
	Green area/inhabitant (SvHab)	4	Agencia d'Ecología Urbana de Barcelona, 2010; Grunewald *et al.* 2017; Istat, 2013; MAES, 2016
	Proximity to green spaces (PV)	4	Agencia d'Ecología Urbana de Barcelona, 2010, EKLIPSE, 2016; Rueda, 2007; Ioja *et al.* 2014
	Structural connectivity	4	Pakzad and Osmond, 2016; Rueda, 2007 ; Ioja *et al.* 2014; Artmann *et al.* 2017
C.7.	Social cohesion	3	Pakzad and Osmond, 2016; EKLIPSE, 2016; Artmann *et al.* 2017

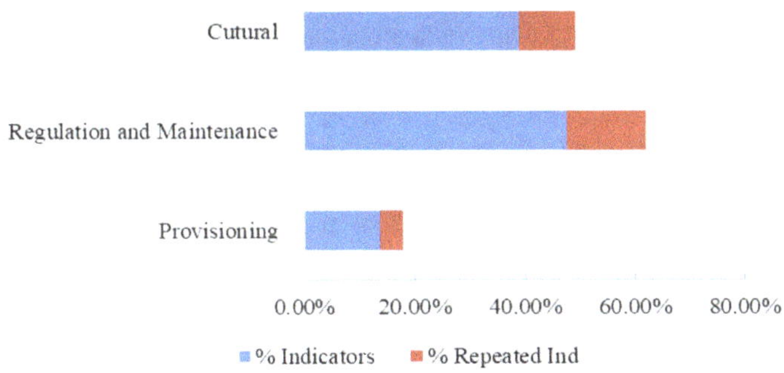

Figure 1: Percentages occupied by indicators for each theme of ES and percentage of repeated ones.

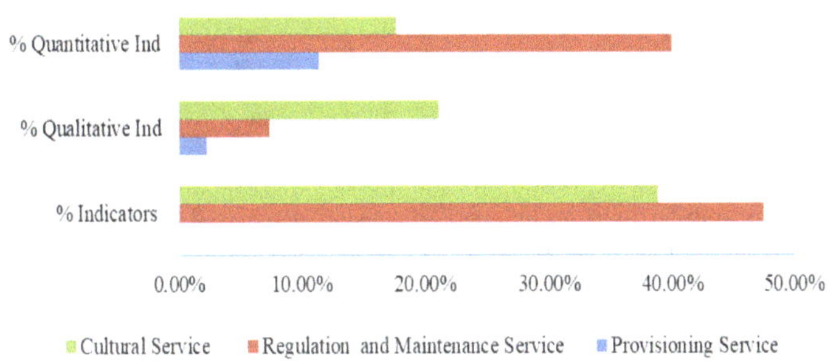

Figure 2: Percentage of qualitative/quantitative indicators associated with ES.

biodiversity [15], mental health changes, opportunities for recreation, or tourism and social interaction (community livability).

Very few works [14–16] take into account the whole range of services provided by urban green infrastructure.

Pakzad and Osmond [16] proposed an integrate method called weighted average index (WAI) based on 16 selected indicators. Agencia d´Ecologia Urbana de Barcelona [15] uses the same type of approach, involving multiple areas without reaching to build an integrated index. Through the current report [15], which we may say that is an ecosystem vision of sustainable city trends, Vitoria-Gasteiz (Spain) sustainable development goals are described and evaluated. Within the green capital, urban indicators assessed urban sustainability in order to create a desirable city model. Agencia d´Ecologia Urbana de Barcelona developed a thematic classification of 50 indicators structured in eight areas that is applied in Vitoria-Gasteiz. Vitoria-Gasteiz Sustainability Plan is representative not only for the holistic vision but also for its practically results after applying urban green indicators and establishing even minimum and desirable objectives.

Several authors [14], [16], [31] wanted to attest the potential of green infrastructure as a planning tool. They are taking into account concepts like social justice or social cohesion, looking for benefits for whole society when it comes to embrace UGI as a new planning tool.

The main purpose of most of the articles, which investigate GI indicators, is to integrate a management plan adopted by policy-makers or stakeholders. Some of the reviewed works [14], [16], [31] emphasize the potential of GI Indicators as a planning tool. In this way, indicators allow establishing goals, estimating the performance, synthesizing and conveying information. The main purpose of using indicators in a policy context is to provide messages to stakeholders and policy actors to achieve better governance and simplified information and communication [37].

None of the reviewed works analyze the interdependence between social, environmental and economic in an explicit way. However, these connections are inferred from the fact that some indicators could measure services of different groups (e.g. total green area, structural connectivity and distribution of green spaces). In this way, further research should explore the potential of co-benefits among different ES. Besides, further research regarding provisioning services (nutrition-marine plan and animals, materials-biotic materials, energy-renewable biofuels), regulation and maintenance services (regulation of waste-bioremediation, flow regulation-mass flow regulation, regulated of biotic environment-pest and disease control) and cultural (symbolic-spiritual) is required to develop service classes and groups indicators to fill the gaps to estimate their benefits.

Urban green infrastructure aims to bring together all three sustainable development spheres, achieve sustainability goals, ensure a healthy community preoccupied for physical, environment, biodiversity conservation and well-being of inhabitants.

4 CONCLUSIONS

The article reviews the most relevant urban green indicators at the urban scale. They are suitable to provide sustainability evidence, resulting also as an effective tool to assess benefits provided by urban green spaces. Green indicators are relevant to the management of green infrastructure and the assessment of its properties. In addition, indicators are a useful tool to communicate stakeholders the benefits of green spaces in urban areas.

This paper classifies urban green indicators according to existing ecosystem and landscape services frameworks.

It is necessary to use urban green indicators to establish the synergies between urban green elements. We should explore indicators that are more complex and that could be useful to interpret the ecosystem services co-benefits. In the existing literature, these interconnections are poorly described and further applied research would be necessary.

It is essential to unify indicators, to create integrated indexes and provide standards with the aid of green indicators, to help the cities to develop an effective urban planning and management tool. Cities sustainability depends on minimum desirable environmental, social and economic objectives and the holistic approach provided by integrated indexes done on proposed indicators. However, we are still far from their incorporation in applied urban planning and management.

ACKNOWLEDGEMENTS

This work was supported by the European Union's Horizon 2020 research and innovation programme under the project Green Cities For Climate and Water Resilience, Sustainable Economic Growth, Healthy Citizens and Environments with reference 730283.

REFERENCES

[1] United States Environmental Protection Agency, A Framework for Sustainability Indicators at EPA, EPA/600/R/12/687, 2012. www.epa.gov/ord (accessed online 15 May, 2018).

[2] United Nations Commission on Sustainable Development, Framing Sustainable Development. The Brundtland Report – 20 Years On, Sustainable Development in action, April 2007. www.un.org/esa/sustdev/csd/csd15/media/backgrounder_brundtland.pdf (accessed on 03 June, 2018).

[3] United Nations, Transforming our world: the 2030 Agenda for Sustainable Development. A/RES/70/1, Seventieth session, Distr.: General, 21 October, 2015, www.sustainabledevelopment.un.org (accessed on 03 June, 2018).

[4] United Nations. New Urban Agenda. United Nations Conference on Housing and Sustainable Urban Development Conference (Habitat III), Quito, Ecuador, 17–20 October, 2017. www.habitat3.org/the-new-urban-agenda/ (accessed on 05 June, 2018).

[5] Hansen, R. & Pauleit, S., From multifunctionality to multiple ecosystem services? A conceptual framework for multifunctionality in green infrastructure planning for urban areas. *Ambio*, **43**, 516–529, 2014.

[6] United Nations Development Programme (UNDP). Sustainable Development Goals. www.undp.org/content/undp/en/home/sustainable-development-goals.html (accessed on 16 July, 2018).

[7] European Commission, Communication from the commission to the European Parliament The Council, The Council, The European Committee of the regions, Green Infrastructure (GI) Enhancing Europe's Natural Capital, Brussels, 2013. ww.ec.europa.eu/environment/nature/ecosystems/docs/green_infrastructures/1_EN_ACT_part1_v5.pdf (accessed on 30 May, 2018).

[8] European Commission. Environment. Nature and Biodiversity. Nature, 2000._ec.europa.eu/environment/nature/natura2000/index_en.htm (accessed on 20 February, 2018).

[9] Valles-Planells, M., Galiana, F. & Van Eetvelde, V., The principles of sustainability are associated to landscape management. *Ecology and Society*, **19(1)**, 44, 2014.

[10] Benedict, M.A. & McMahon, E.T and The Conservation Fund., Green Infrastructure: Linking Landscapes and Communities, Island Press, 2012.

[11] Countryside Agency, Countryside In and Around Towns: The Green Infrastructure of Yorkshire and the Humber. CA, Leeds, 2006. www.leedsbeckett.ac.uk /as/cudem/CIAT_final.pdf (accessed on 10 July, 2018).

[12] Haines-Young, R. & Potschin, M., Common International Classification of Ecosystem Services (CICES) V5.1 Guidance on the Application of the Revised Structure, UK, 2017.www.seea.un.org/sites/seea.un.org/files/lg23_cices_v5.1_final_revised_guidance_03-10-2017.pdf (accessed on 14 February, 2018).

[13] European Environmental Agency. www.eea.europa.eu/ (accessed on 14 February, 2018).

[14] European Union. Horizon 2020. EKLIPSE Expert Working Group on Nature-based Solutions, An impact evaluation framework to guide the evaluation of nature-based solutions projects. (Draft) 19, 2016. www.eklipse-mechanism.eu/apps/Eklipse_data/website/EKLIPSE_Report1-NBS_FINAL_Complete-08022017_LowRes_4Web.pdf (accessed on 10 February, 2018).

[15] Agencia d'Ecología Urbana de Barcelona, Plan de indicadores de sostenibilidad urbana de Vitoria-Gasteiz. Ayuntamiento de Vitoria-Gasteiz, 2010. www.vitoria-gasteiz.org/wb021/http/contenidosEstaticos/adjuntos/es/89/14/38914.pdf (accessed on 10 March, 2018).

[16] Pakzad, P. & Osmond, P., Developing a sustainability indicator set for measuring green infrastructure performance. *Procedia – Social and Behavioral Sciences*, **216**, 68–79, 2016. doi:10.1016/j.sbspro.2015.12.009.

[17] Schyns, J.F., Hoekstra, A.Y. & Booij, M.J., Review and classification of indicators of green water availability and scarcity. *Hydrol. Earth Syst. Sci.*, **19**, pp. 4581–4608, 2015. doi:10.5194/hess-19-4581-201.

[18] Whitford, V., Enos, A.R. & Handley, J.F., City form and natural process-indicators for the ecological performance of urban areas and their application to Merseyside. *Lanscape and Urban Planning*, UK, **57**, pp. 91–103, 2001. Journal ISSN: 0169-2046.

[19] Gregory, R.D., Gibbons, D.W. & Donald, P.F., Bird census and survey techniques. In *Bird Ecology and Conservation, A Handbook of Techniques*, eds. W.J. Sutherland, I. Newton, & R.E. Green, 17–55, Oxford University Press: Oxford, 2004.

[20] Environmental Commission City of Bloomington. Bloomington Environmental Quality Indicators. Fischer, B.C., Steinhoff, M., Mincey, S. & Dye, L., The 2007 Bloomington Street Tree Report 01-07: An Analysis of Demographics and Ecosystem Services, Indiana, 2007.

[21] Rueda, S., El urbanismo ecológico – Un nuevo urbanismo para abordar los retos de la sociedad actual. *Neutra*, ISSN 1138-1507, **5**, 30–37, 2007.

[22] Barrico, L., Azul, M.A., Morais, M.C., Coutinho, A.P., Freitas, H. & Castro, P., Biodiversity in urban ecosystems: Plants and macromycetes as indicators for conservation planning in the city of Coimbra. *Landscape and Urban Planning*, **106(1)**, 88–102, 2012. https://doi.org/10.1016/j.landurbplan.2012.02.011

[23] Cohen, M., Baudoin, R., Palibrk, M., Persyn, N. & Rhein, C., Urban biodiversity and social inequalities in built-up cities: New evidences, next questions. The example of Paris, France. *Landscape and Urban Planning*, **106(3)**, 277–287, 2012. https://doi.org/10.1016/j.landurbplan.2012.03.007

[24] Krasny, M.E., Lundholm, C., Shava, S., Lee, E. & Hiromi Kobori, H., Chapter 30 Urban Landscapes as Learning Arenas for Biodiversity and Ecosystem Services Management. *Urbanization, Biodiversity and Ecosystem Services: Challenges and Opportunities*, 629–644, 2013. https://doi.org/10.1007/978-94-007-7088-1_30

[25] I. Stat, Italian Statistics. Green and other environmental data.Green areas www.dati.istat.it/?lang=en (accessed on 15 May, 2018).

[26] Iojă, C.I., Grădinaru, S.R., Onose, D.A., Vânau, G.O. & Tudor, A.C., The potential of school green areas to improve urban green connectivity and multifunctionality. *Urban Forestry & Urban Greening*, **13(4)**, 704–713, 2014. https://doi.org/10.1016/j.ufug.2014.07.002

[27] Baró, F., Urban Green Infrastructure: Modeling and mapping ecosystem services for sustainable planning and management in and around cities, Thesis for: Ph.D. Program in Environmental Science and Technology (Autonomous University of Barcelona – UAB), 2016. doi:10.13140/RG.2.2.35758.92485/1

[28] European Union, Mapping and Assessment of Ecosystems and their Services, 3rd Report, Mapping and assessing the condition of Europe's ecosystems: Progress and challenges MAES, Mapping and Assessment of Ecosystems and their Services, 2016.

[29] Kabisch, N., Strohbach, M., Haase, D. & Kronenberg, J., Urban green space availability in European cities. *Ecological Indicators*, **70** 586–596, 2016. https://doi.org/10.1016/j.ecolind.2016.02.029

[30] Leff, M., The Sustainable Urban Forest a Step-by-Step Approach, Davey Institute/USDA Forest Service USFS Philadelphia Field Station, 2016. www.itreetools.org/resources/content/Sustainable_Urban_Forest_Guide_14Nov2016.pdf (accessed on 15 March, 2018).

[31] Artmann, M., Kohler, M., Meinela, G., Ganb, J. & Ioan-Cristian Ioja, I.C., How smart growth and green infrastructure can mutually support each other – A conceptual framework for compact and green cities. *Ecological Indicators*, 2017. doi:10.1016/j.ecolind.2017.07.001

[32] Grunewald, K., Richter, B., Meinel, G., Herold H. & Syrbe, R.-U., Proposal of indicators regarding the provision and accessibility of green spaces for assessing

the ecosystem service "recreation in the city" in Germany. *International Journal of Biodiversity Science, Ecosystem Services & Management*, **13(2)**, 26–39, 2017. https://doi.org/10.1080/21513732.2017.1283361

[33] Lundh, J., Indicators for ecosystem services in urban green space management, UPTEC W17 023 Examensarbete 30 hp, 2017. www.w-program.nu/filer/exjobb/Johan_Lundh.pdf (accessed on 1 June, 2018).

USERS' PERCEPTIONS AND ATTITUDES TOWARDS EDIBLE CAMPUS

T.-I. LEE, Y.-H. CHOU & T.-N. HUANG
Department of Horticultural Science, National Chiayi University, Taiwan.

ABSTRACT

Urban agriculture has been regarded as a strategy towards sustainable urban development. It can transform abandoned or underutilized public or semi-public lands into productive space with landscape plants as well as edible crops. Similar to other green space, the edible landscape can provide a range of ecosystem services. However, it could also encounter vandalism and require more inputs into the maintenance, such as irrigation, fertilizing, weeding, pest control, pruning, and harvesting. Thus, the way in which how users perceive edible landscape can affect the delivery of the concept. Among the edible landscape, edible campus is one of the most popular types. They are semi-public space, which surrounds public institutions and is open to the public while imposes a set of stricter rules on users' behaviours than outside. Given the qualities, edible landscape may be more likely to be delivered and maintained in such a semi-public space than in a public space. In Taiwan, edible campuses are often at an elementary school level. Colleges and universities, on the other hand, have their own farms to support agriculture-related practicums and thus pay less attention to the creation of edible campus. However, campuses at a college or university level tend to be more open to the public than at other educational level. Given its semi-public quality, they serve as a pertinent case to explore how users view edible landscape. Through survey, this study collects 406 data from onsite distribution between March and April in 2017 in National Chiayi University, Taiwan. The results reveal that although the concept is unfamiliar to most of the users, they are willing to support the idea of edible campus. In addition, the findings show that users' positive knowledge of edible landscape has association with their support for edible landscape. However, the users' knowledge perception does not have association with their support.

Keywords: edible landscape, edible school, maintenance, productive space

1 INTRODUCTION

Edible landscapes are defined as spaces greened and beautified by using edible crops and/or traditional landscape plants. As a type of green infrastructure components, edible landscapes are increasingly perceived as, not only productive spaces contributing to food security and social well-being [1], but also recreational spaces having potential to beautify underutilized sites, foster social cohesion and community development [2], and increase biodiversity and provide ecosystem services [3].

Given the multiple functions and values, edible landscapes are addressed in an urban or peri-urban more often, than in a rural context, as an approach to sustainable urban development or regeneration. In comparison with farmlands or other urban green spaces, the term suggests it pays more attention to both productive and aesthetic quality. Sometimes, urban agriculture is a synonym for edible landscape as being referred to a range of edible spaces, including green walls, rooftop gardens, back yards, roadside greens, community gardens, allotments, urban parks, farmlands, nursery, and greenhouses [4]. However, urban agriculture can be defined as a term with a wider range of urban production activities, including crops, animals, and even non-food products, such as aromatic and medicinal herbs, ornamental plants, and tree products [5].

The edible landscape in urban public lands may concern visual aesthetics, crop harvest, as well as the needs and limitations of an urban environment. The feature concerning multiple

DOI: 10.2495/DNE-V14-N1-30-40

facets can complicate the establishment and management of edible landscapes in urban public lands. For example, vandalism is a challenge concerned when the Edible Campus project was developed at the McGill University in downtown Montreal [6]. In addition, Seattle's urban forest management plan excludes gathering as a legitimate activity; thus, the management encounters conflicts when fruit harvesting groups regard the forest as a source of food with their roots in the food security movement [7].

Through developing design techniques, neighbourhoods can create edible landscape with decorative, productive, and other functions in a densely built urban environment. Among these edible landscapes, edible campuses are a common type which helps to demonstrate how edible landscape can be woven into urban spaces [6] and connected to the sustainability of urban development [8]. For instance, Bhatt et al. [6] reported the Edible Campus project turning underutilized urban spaces into productive places by incorporating ecological containers to grow crops on paved areas on the downtown campus of McGill University (Fig. 1). However, previous studies focusing on edible campus are at elementary or middle school level and thus pay attention to the educational effects of these garden programs on, for example, students' constructs of life skills [9], or their knowledge, attitudes, and behaviours associated with vegetable consumption [10].

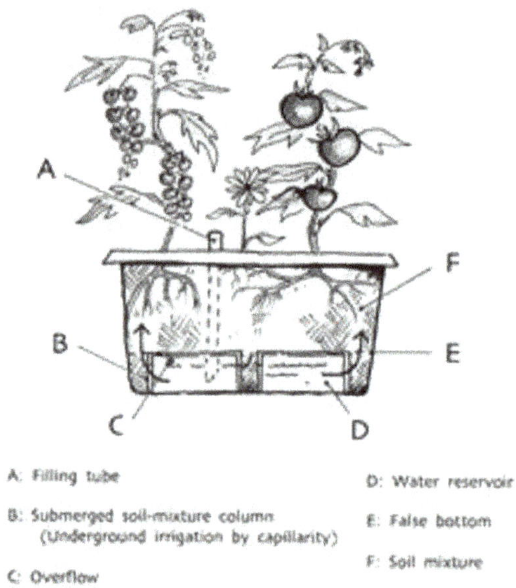

(a)

(b)

Figure 1: Edible Campus projects on McGill University's downtown campus. (a) The ecological containers; (b) photos of the edible campus.
Source: [6].

Although there are cases found worldwide transforming public open spaces in downtown areas, such as Lafayette Greens, Detroit, into edible ones [11], edible plazas as an alternative to traditional greening of public open spaces are still rare. Some studies have focused on how the existing or potential users perceive edible landscape in Western culture. For instance, Sany-Mengual et al. uses semi-structured interviews with stakeholders in Barcelona, Spain and discovered that urban agriculture is largely perceived as a social activity rather than a food production initiative [12]. However, little research has been conducted to analyse the attitudes of potential users of edible plazas in the Eastern Countries. To address these knowledge gaps, this paper examines the main research question: What are the potential users' attitudes towards the hypothetical establishment and management of edible landscaping in underutilized campus plazas as an alternative to the existing greening? In addition, it analyses whether users' supports to the concept of the edible plazas show differences among indexes of their backgrounds, perceptions of edible landscape, and attitudes towards urban agriculture.

2 RESEARCH DESIGN

2.1 Study area

The National Chiayi University in Chiayi, Taiwan (120.27°E, 23.29°N) is selected as the study area. The National Chiayi University has a population over 13,903 people, including visitors, 12,545 students, 497 faculties, and 861 staffs based on the statistics of the university [13]. It has four campuses scattered at different locations in and around the city. They are Lantan Campus, Minsyong Campus, Linsen Campus, and Sinmin Campus (Fig. 2) and each has its own central libraries. In front of the four libraries, there are plazas with traditional greening rather than edible landscaping. The four plazas in front of the central libraries of each campus (Fig. 2) are selected because they tend to be paved open spaces, where movable crop planters can be easily placed. They are underutilized and, thus, subject to revitalization. In addition, they are highly visible and recognizable to the campus users, so the respondents can easily recall how the places look like.

2.2 Questionnaires

A questionnaire is designed to investigate what campus users including staff, faculties, undergraduates, graduates, and visitors perceive the establishment and management of edible plazas on a hypothetical situation, in which edible landscape replaces the existing landscape of the underutilized plaza in the National Chiayi University, Taiwan. The survey was conducted from March 2017 to April 2017. The questionnaires were distributed to users of those busy places on campus during noontime, such as school restaurants, sport fields, and libraries. Questionnaires are disseminated on both working days or non-working days in order to collect data from a wide range of campus users with different purposes, i.e. study, work, or recreation.

We divided the questionnaire into three sections: first, the demographic characteristics backgrounds of the respondents, their planting experience, and their place of residence; second, attitudes towards edible campus (Fig. 1) using ecological containers [6] in replace with the existing landscape of the paved plaza (Fig. 2); finally, opinions on the establishment and management of the edible plaza.

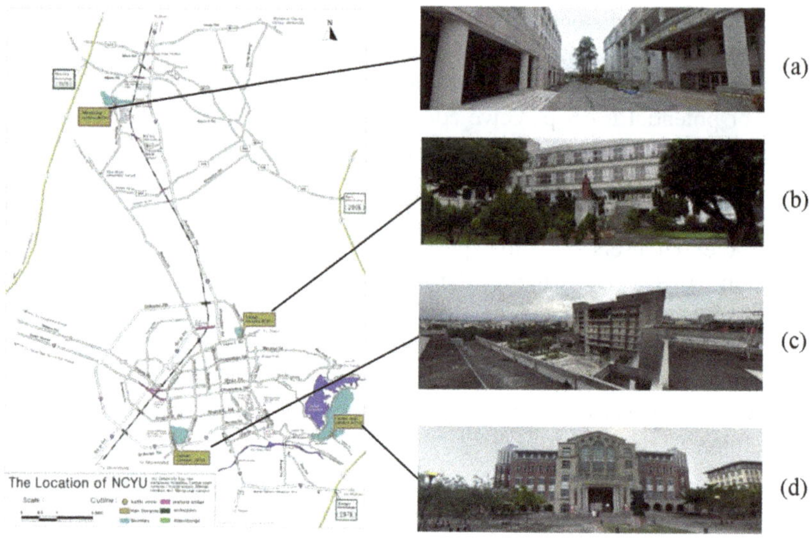

Figure 2: Location map of the four plazas in front of the central libraries in National Chiayi University campuses. (a) Campus Lantan; (b) Campus Minsyong; (c) Campus Linsen; (d) Campus Sinmin.
Source: [13,14].

3 RESULTS

3.1 Basic information of respondents

A total of 406 campus users were surveyed face-to-face on the four campus users randomly with the questionnaire, accounting for about 29.2% of the overall population of the study. The basic information for all respondents is presented (Table 1). This sample consists of 234 women (57.6%) and 172 men (42.4%), roughly indicating the area's gender balance. The age groups 2630 and 3140 were represented in same proportions: 4.4% respectively, which was slightly higher than the proportion (1.2%) of respondents over 41 years old. No respondents indicated that they were less than 18 years of age.

Respondents indicated their locations on the Campus Lantan accounting for 52.7%, Minsyong accounting for 22.7%, Linsen accounting for 3.2%, and Sinmin accounting for 21.4%. On education background, the survey participants were mainly (89.4%) undergraduates, while people who had other level of educational backgrounds were much lower: junior high school education (0.2%), senior high school education (3.4%), and graduate education (7.1%). Regarding the position, 89.4% of the respondents were students, 3.9% were faculties and staff, and 6.7% were visitors. Regarding the age of respondents, the majority was between 18 and 25 years old (89.9%).

Regarding the residence, most respondents lived in urban areas (55.7%), while respondents lived in rural and suburban areas represented in roughly same proportions: 20.2 and 24.1%, respectively. In addition, although the majority of respondents (72.9%) did not have farming training, the majority of respondents (63.1%) had farming experience.

Table 1 also shows the campus users' familiarity and recognition of edible landscaping. Although most people (72%) indicated their unfamiliarity with edible landscaping, a higher proportion (96%) of the respondents agreed with the presence of agricultural activities in urban areas.

Table 1: Basic information of respondents (*N*=406).

Item	Group	Number	Proportion (%)
Gender	Male	172	42.4
	Female	234	57.6
Age	<18	0	0
	18–25	365	89.9
	26–30	18	4.4
	31–40	18	4.4
	41–65	4	1.0
	>66	1	0.2
Campus	Lantan	214	52.7
	Minsyong	92	22.7
	Linsen	13	3.2
	Sinmin	87	21.4
Education	Junior high	1	0.2
	Senior high	14	3.4
	Undergraduate	362	89.2
	Graduate	29	7.1
Position	Student	363	89.4
	Faculty	5	1.2
	Staff	11	2.7
	Visitor	27	6.7
Residence	Rural	82	20.2
	Suburban	98	24.1
	Urban	226	55.7
Agricultural training	Yes	110	27.1
	No	296	72.9
Farming experience	Yes	256	63.1
	No	150	36.9
Familiarity with the term	Yes	113	28
	No	293	72
Urban agriculture	Agree	391	96
	Not agree	15	4

3.2 Subjective knowledge towards edible plazas

In comparison with traditional landscaping and greening, we investigate the campus users' subjective knowledge towards edible plazas. Each item was evaluated individually on a five-point Likert-scale. Table 2 displays the mean and standard deviation. Results show that campus users feel edible plazas need more maintenance than the traditional landscape. In addition, they think edible plazas are more educative, attractive, sustainable, beautiful, and providing more functions than the traditional ones. However, edible plazas are also regarded as being more subject to vandalism. In addition, the mean of the statements that edible plazas

Table 2: Subjective knowledge towards edible plazas (*N*=406).

Item	Mean	Median	SD
Need more maintenance	4.14	4	0.850
More educative	4.10	4	0.852
Provide more functions	3.86	4	0.852
More attractive	3.83	4	0.944
More subject to vandalism	3.81	4	0.971
More sustainable	3.69	4	1.003
More beautiful	3.40	3	0.973
Consume more energy	2.97	3	1.068
Smellier	2.95	3	1.022

Note: SD=standard deviation.

consume more energy and are smellier than traditional landscaping are below the midpoint three indicating general disagreement.

3.3 Attitudes towards the management of edible plazas

We then investigated attitudes towards the maintenance of edible plazas. Each item was evaluated individually on a five-point Likert-scale. Results in Table 3 show that regarding maintenance, chemical compounds prohibited, social activities held on a regular basis, and selling the harvests are most valued management by the campus users. However, the mean of the statement harvesting activities open to the public is below the midpoint three indicating general disagreement.

3.4 General attitudes towards of edible plazas

Afterwards, we investigated campus users' attitudes towards edible plazas (Table 4). Each item was evaluated individually on a five-point Likert-scale. Campus users agree the most with the ideas that edible plazas are a form of campus greening as well as a form of urban agriculture. In addition, they think edible plazas are appropriate and show willingness to support the concept of edible plazas. However, campus users show less agreement on the susceptibility to contamination and the nontoxicity of the edible plazas. Although they show

Table 3: Managerial attitudes towards edible plazas (*N*=406).

Item	Mean	Median	SD
Chemical compounds prohibited	4.04	4	0.979
Social activities held on a regular basis	3.92	4	0.579
Selling the harvests	3.92	4	0.859
Managing the harvests	3.83	4	0.936
Harvesting open to the public	2.88	3	1.099

Note: SD=standard deviation.

Table 4: General attitudes towards edible plazas (N=406).

Item	Mean	Median	SD
It is a form of campus greening	4.06	4	0.775
I am willing to support it	3.99	4	0.815
It is a form of urban agriculture	3.96	4	0.782
It is appropriate	3.86	4	0.797
I will eat the harvests	3.80	4	0.896
It is subject to contamination	3.67	3	1.047
It is nontoxic	3.57	4	0.883
I will pick up the harvests	3.46	4	1.057

Note: SD=standard deviation.

more willingness to eat the harvest, they show less willingness to pick up the harvests. Overall, the mean of the statements are above the midpoint three indicating general agreement.

The data were analysed using exploratory factor analysis. Table 5 shows the rotated component matrix. Through the analyses, seven were deleted from the 22 questions due to a loading value below 0.5 or a cross-loading issue. The KaiserMeyerOlkin (KMO) criterion is 0.821, which is considered to be meritorious. Through the factor analysis, the following five factors were found to indicate campus users' attitudes:

Factor 1 (F1): Merits of edible plazas
Contains positive items which express that edible plazas would be more sustainable, attractive, educative, and beautiful and they would provide more functions than the existing plazas. The Cronbach's alpha measures 0.802, which is considered to be good.

Factor 2 (F2): Food safety issues of edible plazas
Sums up items in relation to food safety which express that the campus users would pick up, eat the harvests from crops grown on the edible plazas, and the harvests would be nontoxic. The Cronbach's alpha measures 0.734, which is considered to be acceptable.

Factor 3 (F3): Demerits of edible plazas
Contains negative items which express that edible plazas would be smellier, subject to contamination, and they would consume more energy. The Cronbach's alpha measures 0.705, which is considered to be acceptable.

Factor 4 (F4): Support for edible plazas
Sums up statements concerning support for the edible plazas, such as the recognition of the edible plazas as a form of urban agriculture and campus greening and the willingness to support the edible plazas. The Cronbach's alpha measures 0.738, which is considered to be acceptable.

Factor 5 (F5): Management of edible plazas
Contains merely a managerial item expressing that harvesting the edible parts of the crops grown on the edible plazas should be open to the public.

3.5 Perception and attitude determinants of support for edible plazas

The analysis tests how factors and indexes of perceptions and attitudes affect campus users' intentions towards for edible plazas. Table 6 presents results from Spearman's correlation coefficient Rho measures, investigating the influence of merits, demerits, food safety,

Table 5: Attitudes rotated factor loading matrix (N=406).

	Factor 1	Factor 2	Factor 3	Factor 4	Factor 5	Communality
KMO: 0.821	Merit	Safety	Demerit	Support	Manage	
Sum of squared loading	4.488	1.964	1.349	1.057	1.012	
Percentage of variance explained (%)	29.920	13.095	8.991	7.056	6.747	
Cumulative percentage of variance	29.920	43.016	52.006	59.062	65.809	
Cronbach's alpha	**0.802**	0.734	0.705	0.738	-	
The edible plaza would be more sustainable	**0.763**	0.232	−0.098	0.077	0.005	0.651
The edible plaza would be more attractive	**0.735**	0.130	0.109	0.198	0.120	0.623
The edible plaza would be more educative	**0.702**	0.119	−0.013	0.067	−0.032	0.512
The edible plaza would be more beautiful	**0.701**	0.051	0.053	0.175	0.028	0.529
The edible plaza would provide more functions	**0.682**	0.136	0.120	0.216	−0.043	0.546
I would pick up the harvests	0.216	**0.822**	−0.065	0.123	−0.048	0.749
I would eat the harvests	0.200	**0.817**	0.070	0.063	0.178	0.744
The harvests would be nontoxic	0.115	**0.681**	0.049	0.190	−0.131	0.532
The edible plaza would be smellier	−0.008	0.079	**0.809**	−0.045	−0.118	0.676
The edible plaza would consume more energy	0.016	0.026	**0.807**	0.167	0.029	0.681
The harvest would be subject to contamination	0.087	−0.062	**0.744**	−0.093	0.097	0.584
The edible plaza would be a form of campus greening	0.176	0.198	−0.094	**0.817**	0.049	0.749
The edible plaza would be a form of urban agriculture	0.265	0.069	0.173	**0.762**	−0.016	0.686
I am willing to support the edible plaza	0.463	0.318	−0.142	**0.566**	0.048	0.658
Harvesting of the edible plaza should be open to the public	0.033	−0.016	0.011	0.041	**0.974**	0.952

Note: KMO=KaiserMeyerOlkin criterion.

management, subjective knowledge, managerial attitude, and general attitude on the support for edible plazas.

Looking at the merit factors (F1), the results show that a general recognition of the merits of edible landscapes increases support for the edible campus. However, the demerit factors (F3) have no significant impact on the support of edible campus. Furthermore, the food safety factors (F2) have a significant and positive effect on campus users' support for the

Table 6: Perception and attitude determinants of support for edible campus (N=406).

	F4	I am willing to support the edible plaza
Items	rs	rs
F1: Merits	0.529*	0.510*
F2: Food safety	0.422*	0.437*
F3: Demerits	−0.28	−0.084
F5: Management	0.69	0.44
Subjective knowledge index	0.425*	0.369*
Managerial attitude index	0.374*	0.270*
General attitude index	0.773*	0.710*

edible plaza. In addition, the indexes of subjective knowledge, managerial attitude, and general attitude have a significant and positive influence on campus users' support for the edible plazas.

3.6 Background determinants of support for edible campus

The analysis is completed by examining how indexes of background affect campus users' intentions towards for edible plazas. Table 7 presents results from t-test and one-way ANOVA (analysis of variance) measures, investigating the influence of gender, age, campus, education, position, residence, agricultural training, farming experience, familiarity with edible campus, and recognition to urban agriculture on the support for edible plazas.

For the indexes of gender, agricultural training, farming experience, familiarity with edible campus, and urban agriculture, the results of t-test present that these indexes do not have significant effect on campus users' support for the edible plaza or F4.

Table 7: Background determinants of support for edible campus (N=406).

Item	Support		F4	
	F	P	F	P
Gender	0.413	0.521	2.026	0.155
Agricultural training	3.861	0.050	5.044	0.025
Farming experience	0.402	0.526	0.006	0.937
Familiarity with edible campus	0.109	0.742	0.652	0.420
Urban agriculture	0.244	0.621	0.176	0.675
Age	3.929	0.004*	2.728	0.029
Campus	0.188	0.904	0.620	0.602
Education	5.104	0.002*	2.405	0.067
Position	0.936	0.423	1.457	0.226
Residence	0.661	0.517	1.256	0.286

Regarding the effect of the indexes of age, campus, education, position, and residence on campus users' support for the edible plaza, the *p*-values of all the indexes for the Levene statistic (0.756/p=0.519, 1.823/p=0.142, 1.059/p=0.348, 0.411/p=0.745, and 3.526/p=0.30, respectively) are not significant (p>0.05). Thus, the assumption of homogeneity of variances is tenable. The results of the ANOVA *F*-test present that only the indexes of Age and Education have significant effect on campus users' support for the edible plaza (p<0.01).

Regarding the effect of the indexes of age, campus, education, position, and residence on F4, the *p*-values of all the indexes for the Levene statistic (2.199/p=0.088, 1.127/p=0.338, 2.145/p=0.118, 2.644/p=0.049*, and 0.792/p=0.453, respectively) are not significant (p>0.05). Thus, the assumption of homogeneity of variances is tenable. The results of the ANOVA *F*-test present that no indexes have significant effect on F4 (p<0.01).

4 DISCUSSION AND CONCLUSION

Edible landscape has gradually been recognized as providing a wide range of advantages and contributing to sustainability. However, it may also encounter vandalism, disadvantages, and other maintenance issue. Interestingly, although the campus users agree with advantages of edible plazas, they do not think edible plazas have disadvantages, such as consuming more energy or being smellier than the traditional ones. Given these advantages and disadvantages of edible landscape, this study examines whether users' subjective knowledge affects their supports for the hypothetical transformation of existing campus plazas into edible plazas. The findings show that users' positive knowledge of edible landscape has association with their support for edible landscape. However, the users' knowledge perception does not have association with their support. Similarly, Grebitus et al.'s study reveals that the index which encourage urban farm purchase has significant association with purchase likelihood, but the index prevent urban farm purchase does not have significant association with purchase likelihood [15].

Looking at the background characteristics of users, the users' agricultural experiences or training do not affect their support for edible plaza, whereas age and education affect their support for edible plaza. However, the results may be biased by the data concentration to a certain age and education range as almost 90% of respondents are undergraduate students within the age range between 18 and 25 years old. This forms the limitation of this research. Therefore, future studies can examine users' perception and attitude with a wider range of age, education, or other background characteristics. The hypothetical nature of this study is highlighted. The survey tests campus user's perceptions and attitudes on a hypothetical situation rather than an actual situation of edible plazas. Therefore, future research might include a real, non-hypothetical transformation of public urban places into other types of edible landscape, such as edible playgrounds, edible parks, edible avenues, or edible alleys. Finally, future surveys on willingness to pay for the transformation or the maintenance of the edible places would be beneficial. Overall, this study provides an outlook on the acceptance of edible landscape. It also shows which perception or attitudinal index affects the support for the hypothetical transformation of the existing plazas into edible one on campus.

REFERENCES

[1] Mok, H.F., Williamson, V.G., Grove, J.R., Burry, K., Barker, S.F. & Hamilton, A.J., Strawberry fields forever? Urban agriculture in developed countries: a review. *Agronomy for Sustainable Development,* **34**(1), 21–43, 2014. https://doi.org/10.1007/s13593-013-0156-7

[2] Contesse, M., van Vliet, B.J.M. & Lenhart, J., Is urban agriculture urban green space? A comparison of policy arrangements for urban green space and urban agriculture in Santiago de Chile. *Land Use Policy*, **71**, 566–77, 2018. https://doi.org/10.1016/j.landusepol.2017.11.006

[3] Lin, B.B., Philpott, S.M. & Jha, S., The future of urban agriculture and biodiversity-ecosystem services: challenges and next steps. *Basic and Applied Ecology*, **16(3)**, 189–201, 2015. https://doi.org/10.1016/j.baae.2015.01.005

[4] Cretella, A. & Buenger, M.S., Food as creative city politics in the city of Rotterdam. *Cities*, **51**, 1–10, 2016. https://doi.org/10.1016/j.cities.2015.12.001

[5] Urban agriculture; Food and Agriculture Organization of the United Nations (FAO), http://fao.org/urban-agriculture/en/ (accessed 26 August 2018).

[6] Bhatt, V., Farah, L.M., Luka, N. & Wolfe, J.M., Making the edible campus: a model for food-secure urban revitalisation. *Open House International*, **34(2)**, 81–90, 2009.

[7] McLain, R., Poe, M. & Hurley, P., Lecompte-Mastenbrook, J., Emery, M.R., Producing edible landscapes in Seattle's urban forest. *Urban Forestry & Urban Greening*, **11(2)**, 187–94, 2012. https://doi.org/10.1016/j.ufug.2011.12.002

[8] Fischer, D.B., Karle, S.J., Cremer, K., Huttner, E., Seebauer, M., Nowikow, U., Schütze, B., Voigt, P., Völker, S. & Kowarik, I., Biodiverse edible schools: linking healthy food, school gardens and local urban biodiversity. *Urban Forestry & Urban Greening*, in press.

[9] Robinson, C.W. & Zajicek, J.M., Growing minds: the effects of a one-year school garden program on six constructs of life skills of elementary school children. *HortTechnology*, **15(3)**, 453–57, 2005.

[10] Ratcliffe, M.M., Merrigan, K.A., Rogers, B.L. & Goldberg, J.P., The effects of school garden experiences on middle school-aged students' knowledge, attitudes, and behaviors associated with vegetable consumption. *Health Promotion Practice*, **12(1)**, 36–43, 2011. https://doi.org/10.1177/1524839909349182

[11] Philips, A., *Designing Urban Agriculture: A Complete Guide to the Planning, Design, Construction, Maintenance and Management of Edible Landscapes*, Wiley: Hoboken, 2013.

[12] Sanyé-Mengual, E., Orsini, F., Oliver-Solà, J., Rieradevall, J., Montero, J.I. & Gianquinto, G., Techniques and crops for efficient rooftop gardens in Bologna, Italy. *Agronomy for Sustainable Development*, **35(4)**, 1477–88, 2015. https://doi.org/10.1007/s13593-015-0331-0

[13] Map of School Campus. National Chiayi University, http://ncyu.edu.tw/eng/content.aspx?site_content_sn=36655 (accessed 26th August 2018).

[14] Google Street View, https://google.com.tw/maps/ (accessed 26 August 2018).

[15] Grebitus, C., Printezis, I. & Printezis, A., Relationship between consumer behavior and success of urban agriculture. *Ecological Economics*, **136**, pp. 189–200, 2017. https://doi.org/10.1016/j.ecolecon.2017.02.010

PRODUCTION AND ECONOMIC ANALYSIS OF A POP-UP FARM IN MEXICO CITY

P.D.H. GUERRERO[1], L.M.M. MATA[1], P.B. MADRIGAL[1] & J.J.M. CRIOLLO[2]
[1] Mechanical Engineering, Engineering Faculty, La Salle University, Mexico.
[2] Solid State Section, Centro de Investigación y de Estudios Avanzados, Mexico.

ABSTRACT

Urban agriculture is becoming a relevant strategy to increase the resilience in highly populated cities since it helps in addressing urban food scarcity, malnourishment and obesity problems, may provide alternative economic opportunities and improves environmental quality. Mexico City is facing numerous challenges related with the fast growth of its population that threatens its ability to keep up providing them with basic food and water services. Low prices of processed foods associated with industrial agriculture, combined with the urbanization lifestyle, have left Mexico City with a public health crisis. To contribute to solving these problems, it is proposed to use several Pop-Up Farms using Mel Bartholomew's system for the cultivation of local vegetables (tomatoes, romaine lettuce, pepper, spinach, onions, celery, etc.), taking advantage of the city's subtropical highland climate and a specially designed rainwater harvesting system. Preliminary results show that a Pop-Up Farm can supply a significant fraction of the average yearly family consumption for these vegetables in the region. A cost benefit analysis shows that Pop-Up Farms are a viable strategy for providing high-quality food at low costs contributing to increase health and economic conditions in urban environments.

Keywords: cultivation table, low cost, Mel Bartholomew's system, rainwater, urban agriculture

1 INTRODUCTION

Urban agriculture is not a recent concept in Mexico City. The first human settlements in the area, the Aztecs, in Teotihuacan, were immersed in a complex agricultural landscape. As a result, they established amazing agriculture methods. The most popular is the one called *chinampas,* still used nowadays [1,2]. In Mexico City, urban agriculture coexists in urban and peri-urban areas between rural and urban environments. For example, peri-urban farmers that work in typical jobs (guards, civil servants, construction workers, etc.) during weekends, work their crop fields. They represent 0.7% of the urban population [3].

A limiting factor for urban and peri-urban agriculture to represent a viable strategy to increase food supply and improve urban conditions is the lack of economic incentives. Mexican policies are not currently focused on the development of national urban agriculture. One of the most important urban agriculture programs in Mexico City is the 'Sustainable Small-scale Agriculture' (ASPE) supported by the Ministry of Rural Development and Equity for Community (SEDEREC is the acronym in Spanish). The City Government organizes this program every year with the objective of giving economic support to projects where self-consumption and commercialization of organic foods at small scale are implemented. In 2017, 131 projects in the 'Promote Urban Agriculture' division were approved [4].

On the other hand, in industrialized countries where food security is not compromised, urban agriculture is viewed more as a recreational exercise than an actual activity for food production. However, in developing countries like many Latin American's it has become a relevant activity for food production [5].

In this paper, we are proposing the use of two Pop-Up Farms cultivation tables in Mexico City, in Gustavo A. Madero Municipality. In this location, like most of the city, we can find

DOI: 10.2495/DNE-V14-N1-41-51

great economic inequalities and underdevelopment. According to the National Institute of Statistics and Geography [6], approximately 10.8% of the people in the area are affected by food scarcity. Additionally, data of urban agricultural production in the zone is missing. As the Food and Agriculture Organization observes, urban agriculture in this part of Mexico City is in its beginning [7].

A significant advantage to maximize production of our Pop-Up Farming cultivation tables is that they follow the empirical Mel Bartholomew's system (MB system) [8]. We build the two elevated cultivation tables according to the technique 'Square meter gardening', in a space of 1 m^2, it is feasible to plant crops that provide fresh vegetables for one person all year round. Following Bartholomew's method, the crops are rotated according to their harvest cycles. Furthermore, the MB system identified, by experimental research, the best mix of soil and the necessary space each crop needs to allow using the whole growing area. We choose the MB system due to its simplicity. It can be operated by a common person (non-agricultural). Finally, to evaluate the economic benefits that represent our production, we have done a cost-benefit analysis which consists of a comparison between the expected results of the MB system with our results.

To our knowledge, the use of the rainwater in Mexico City is rather scarce, despite Mexico having the conditions to exploit this resource given its geographic location and weather. There is a minimal amount of recycled water or rainwater harvest in houses. Furthermore, the drinking water supplies in Mexico City are an increasing problem as a result of the overpopulation, poor urban planning, general contamination and lack of regulations to enforce proper waste water treatment. In the present work, we will address the benefits of implementing a rainwater harvest system for crop irrigation [3].

This paper is organized as follows. Section 2 describes the implementation of the cultivation tables and the design of the rainwater capture system. Section 3 contains the procedure of the costbenefits analysis. In Section 4, we present the performance of the two cultivation tables and their corresponding economic analysis. Section 5 presents discussions and concluding remarks.

2 CONSTRUCTION OF THE CULTIVATION TABLES

This section briefly describes the two cultivation tables used in this work based on the MB system and complemented by an *ad-hoc* rainwater harvesting system.

The cultivation tables were placed following the three main factors of the MB system: (i) it must be easily accessible to the user to facilitate plant care and crop harvest, (ii) it should have between 6 and 8 h of sunshine per day and (iii) a suitable drainage system must be ensured to prevent excess humidity in the cultivation tables' substrate (ground). The location of the cultivation tables is presented in Fig. 1.

We have used the recommended procedures to design and build the two elevated cultivation tables. The MB system suggests that approximately 30 cm depth of soil mixture is suitable to plant all kinds of crops. The cultivation tables were manufactured using common and widely available materials, such as ordinary wood panels, bricks, plastic, nails and screws. The first cultivation table, labelled as E in Fig. 1, is 1 m^2 with 30-cm depth where crops that develop larger roots were placed and the second one, F, is 1.8 m × 1 m with 15-cm depth were adequate crops were planted. Different sizes were used for two main reasons, the material and space we had available. Following the MB method, both cultivation tables were divided into rectangular sections of at least 30 cm per side as indicated in Fig. 2. The area of the sections provides the necessary space for each plant when they are mature. Cultivation

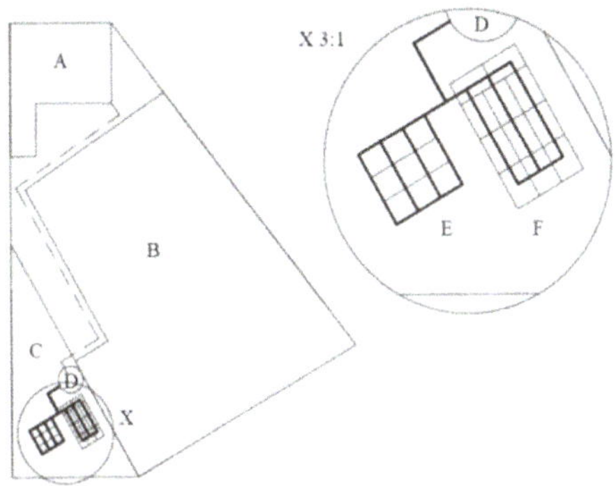

Figure 1: Plan view of the location of the cultivation tables where *A* is the rainwater harvest area (approx. 14 m^2), B is the total area of the house's roof of 100 m^2, C is the location area of the cultivation tables, D is the water tank and E and F are the cultivation tables. The irrigation system is represented by a bold line above E and F and the rainwater harvesting system by a dashed line.

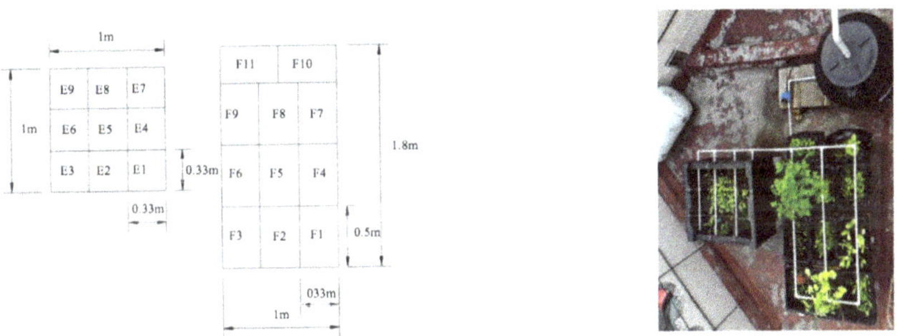

(a)

(b)

Figure 2: Cultivation tables. (a) Identification of sections of both cultivation tables E and F. (b) Photo of the corresponding cultivation tables.

table *E* was divided into nine sections E1E9, while cultivation table *F* was divided into 11 sections F1–F11.

2.1 Crop selection

Mexico City presents an advantage due to the weather conditions which are favourable for flora's development. The climate is mild and moderate to abundant rainfall in summer. The average annual temperature is 16°C, highest temperatures (25°C approx.) are reached from March to May and the lowest (5°C approx.) during January [6].

For this work, a mixture of 12 crops and herbs commonly found in a Mexican diet were selected: onion, celery, coriander, lettuce, arugula, radish, spinach, pumpkin, bean, pea, pepper and tomatoes. This selection allows harvesting all year round due to their cultivation season. Crop type and number of sowed seeds per section are shown in Table 1. In section F11, tobacco and chamomile were cultivated as pest repellent and they are not considered for final results.

Table 1: Number of plants by section of cultivation tables.

Plant	Section	Number of plants per section	Total
Onion (Liliaceae)	E1, E7, E8, E9	3, 4, 5, 5	17
Celery (Umbellate)	F6	8	8
Coriander (Umbellate)	F11	10	10
Lettuce (Composite)	E4, E8, E9, F4, F5	2, 4, 2, 6, 6	20
Radish (Cruciferous)	E5, E6	10, 10	20
Spinach (Chenopodiaceae)	F7	20	20
Pumpkin (Cucurbitaceae)	F2	4	4
Arugula (Composite)	E7	4	4
Bean (Labiates)	F9	16	16
Pea (Solanaceae)	F8	16	16
Pepper (Solanaceae)	F1, F3	2, 2	4
Tomato (Solanaceae)	E2, E3	1, 2	3

2.2 Substrate

The substrate used in the cultivation tables is a mixture of three equal parts of coconut fibre, worm humus and compost, commonly known as Mel Mix [9]. The coconut fibre, used as a component of peat-based substrates, provides a moisture retention capacity (66%), a high aeration of the root system, buffer capacity (pH 5) and electrical conductivity of the medium (2.15 mS/cm) [10]. Compost is characterized by keeping the micro fauna in balance, having low levels of salinization, high cation exchange capacity as well as maintaining the physical structure that allows water and air flowing permanently. Finally, worm humus is a substrate with high microorganism content which allows an increase in the biological activity of the soil that in turn improves crops resistance against plagues and diseases [11].

2.3 Germination and heights of the crops

For the cultivation tables, seeds from a supplier were used in Mexico City, 'Los Molinos' ranch [12]. It was possible to plant the first part of the seeds corresponding to the cultivation table E on July 1st. Due to the delay in obtaining the remaining seeds and the materials for the required mixture of the method, the second part of the seeds which corresponds to the cultivation table F, were planted on July 14.

For the days of germination expected, we rely on the information provided by the seed supplier [12]. Germination days were considered as the days in which we saw leaves bud for the first time. Heights were measured daily and were compared with Mel Bartholomew expected cultivation times [8]. Also, we checked the recommended season of cultivation. The detailed information is presented in Table 2.

2.4 Irrigation system

As a first approach, an irrigation schedule using a total of 30 L of water every 2 days was deemed adequate to maintain optimal humidity conditions in both cultivation tables for the whole experimental period. The irrigation was performed by an automatized *ad-hoc* system

Table 2: Germination date and expected heights.

Our experimental data		'Los Molinos' Ranch [12]		Mel Bartholomew [8]		
Plant	Recorded germ. date	Max. expected germ. days	Min. expected germ. days	Max. expected height (cm)	Min. expected height (cm)	Cultivation season
Onion	July 10th	14	10	30	30	Spring and summer
Celery	August 14th	20	14	40	40	Summer and autumn
Coriander	July 24th	12	8	45	30	Spring and summer
Lettuce (E)	July 8th	10	5	30	15	Summer and autumn
Lettuce (F)	July 22th	10	5	30	15	Summer and autumn
Radish	July 4th	6	4	30	15	Summer and autumn
Spinach	July 24th	12	8	30	15	Spring, autumn and winter
Pumpkin	July 20th	8	6	120	45	Summer
Arugula	July 12th	14	10	25	15	Spring, summer and autumn
Bean	July 18th	6	4	200	150	Summer
Pea	July 20th	6	3	150	60	Spring and Summer
Pepper	July 29th	20	15	60	30	Summer
Tomato	July 26th	14	10	200	200	Summer

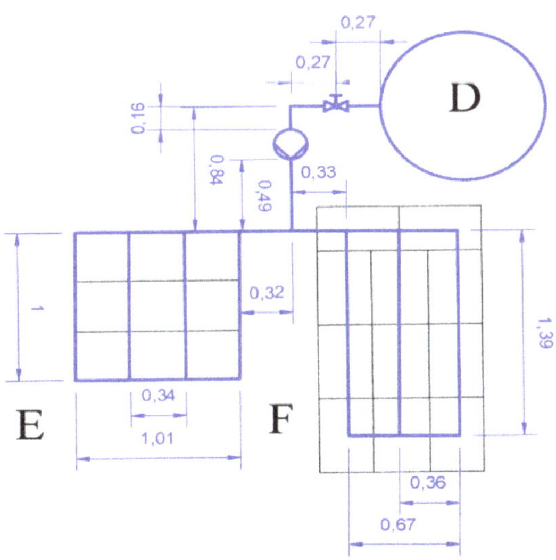

Figure 3: Irrigation system. The blue line represents a set of conduction, valves, PVC connections and distribution pipes installed on the cultivation tables. Here D is the water tank.

that collects rainwater and distributes it with an 1/6 hp water pump. It is estimated that in a 14 m^2 area, it is possible to collect up to 47 m^3 yearly of water given current rainfall conditions in Mexico City easily covering the water needs of the proposed system.

Data including substrate humidity and pH levels (using a Soil Meter sensor), germination period and plant height are logged manually daily. Although not ideal, plant height was used in this work as a measurement of preliminary system performance given that the experiments are ongoing.

3 COST BENEFIT ANALYSIS

A financial model that considers the benefits of the Pop-Up Farm system (cultivation tables, rainwater capture and irrigation systems) and the costs of water, energy usage, implementation, maintenance and operation is developed to evaluate its strengths and weaknesses.

3.1 Categorization of expenses

A list of the expenses incurred during the construction of the proposed production system was recorded to compare and ensure that future designs are more efficient economically and to potentially assess the its lifecycle at the end of the harvest period.

Expenses associated with the construction of the cultivation tables and the rainwater harvest and irrigation systems are considered as implementation costs. Expenses that are directly related to keeping the production system infrastructure in good working order are considered as maintenance costs.

3.2 Financial model

A financial model was built to allow estimations of the expenses that a family might incur to purchase their vegetables and to be able to assign a utility to the vegetables produced by the production system [13,14]. The following data is considered in the model: current inflation rate in Mexico based on monthly consumer price index (CPI) from the past 5 years, current minimum wage in Mexico City; monthly transportation cost (one trip formed by one round subway fare and one round bus fare) incurred to purchase the vegetables [15] in the biggest city market, known as Central de Abastos. It is worth noting that although smaller open markets are common across Mexico City, it is fairly common to purchase family vegetables from the central market due to their lower prices.

In our financial model, the following assumptions were made: inflation rate of 6%, monthly minimum wage ($88.35 MXN) and transport spending ($44.18 MXN) [16].

To estimate a crop's maximum gross financial potential from the production system, the annual production for each crop shown in Fig. 6 is multiplied by the maximum retail price [15] in the market Central de Abastos. In this preliminary study, the prices for August are assumed as constant disregarding seasonal variation in prices. The total gross potential financial gain of production system is simply the total for all crop.

The net financial potential gain of the production system is the estimated gross gain after deducting maintenance expenses. Implementation costs are only considered in the first year because it is the amount of capital required to build the whole system.

3.3 Study case

A study case is proposed to estimate the potential financial benefits that a family might have from growing their own vegetables at their home when compared with the economic cost and time invested in purchasing from the market. For this, the net financial potential, based on central market prices [15], is extrapolated to a 5-year period, assuming prices and expenses increase with inflation rate while system production is assumed constant.

4 RESULTS

This section presents measured system results from germination time, plant height, plant growth rate and crop production. From these, results from the costbenefit analysis are presented.

4.1 Germination time

The measured germination time for each kind of vegetable is presented in Fig. 4. The results are compared with minimum and maximum expected germination times reported by [8,12] and shown in Table 2. Note that the germination times presented for most in the production system are between the expected ones [17]. For other vegetables (e.g. beans, pumpkin and radish), the germination period is close to the minimum expected. This might be related with weather conditions in Mexico City with average annual temperature of 16°C that is beneficial for these crops (seed instructions recommend, for an optimal period of germination, an average annual temperature of 15°C).

4.2 Plant height

The measured plant height after 50 days is presented in Figs. 5 and 6 compared with the expected height values reported by Mel Bartholomew for harvest time [8]. Note that after approximately 45 days, just a few plants have reached harvest time, specifically: coriander and radish. As is clearly seen in Fig. 5, the heights of the plants reached in the proposed production system are lower than the maximum expected heights for all crops: this is expected since the plants have not reached harvest time (only onion, lettuce, radish and spinach reached the minimum expected heights, based on Mel Bartholomew's work [8], within the measured timeframe) by the time reported in Fig. 5.

It is possible to observe that except for coriander and radish, our vegetables are around 50% of the required days, despite being planted in the recommended season presented in Table 2. It is important to mention that the present results are preliminary due to the fact that almost all the crops continue within the growth period to reach their expected height. Even

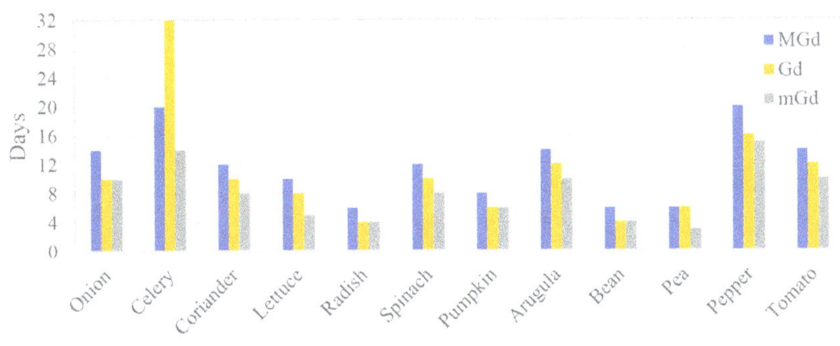

Figure 4: Germination vs days. MGd are the maximum expected germination days, Gd are the germination days presented in our system and mGd are the minimum expected germination days.

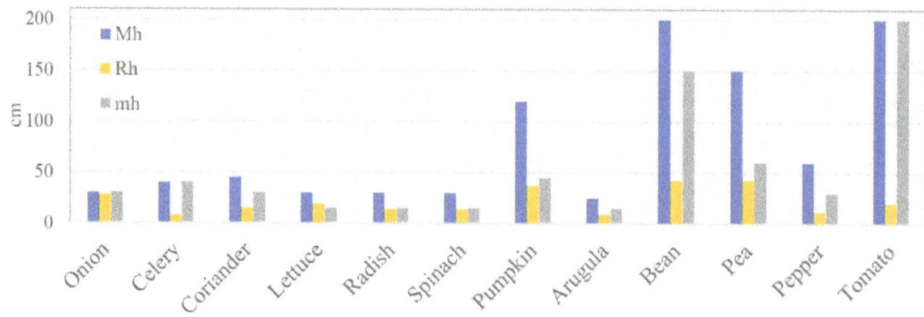

Figure 5: Heights of plants. Mh are the maximum expected heights, Rh are the heights reached by the plants in the proposed production system and mh are the minimum expected heights.

so, we discovered that in the case of the onion, lettuce, spinach and pumpkin, it is possible to obtain at least the minimum expected heights in fewer days.

4.3 Production

In Fig. 6, we display the values estimated for our production. Unfortunately, due to the lack of time, we estimated the production with a rough comparison of time and heights. Of course, such simplifications do impose some limitations on the costanalysis results. Once we had the final production, a natural further step would be to remove the production's estimated values.

With the results obtained in the plant's heights vs. the days elapsed and comparing with the corresponding expected values reported by [8], we calculate the performance per vegetable. The results of this approach are as follows: onion 210.8%, celery 58.3%, coriander 24.3%, lettuce 214.5%, radish 21.1%, spinach 81.7%, pumpkin 54%, arugula 65%, bean 62.7%, pea 58.5%, pepper 72.9% and tomato 28.2%. To use as a reference, 100% is when the vegetable reaches the minimum height within the expected time period reported by [8]. Finally, we multiply the expected production data by the obtained performance's percent and the number of plants in our system.

Our production results using the performance approach shows that crops such as coriander, radish, pumpkin, arugula, bean, pea, pepper and tomato need the complete period of days to

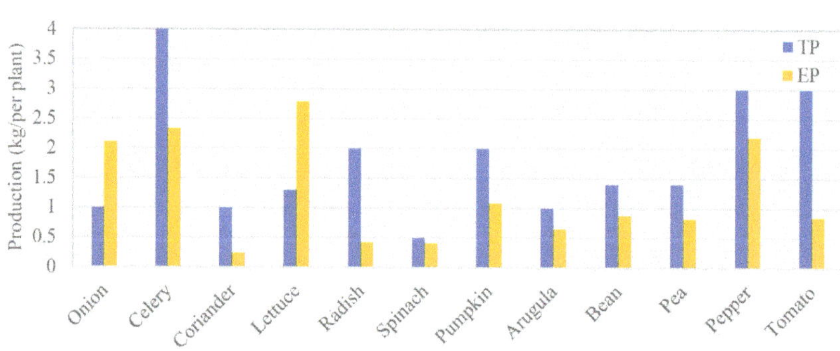

Figure 6: Production per plant. TP are the expected production and EP are the estimated production.

Table 3: Costbenefit results projected for 5 years.

	Year 1	Year 2	Year 3	Year 4	Year 5
Income (in $)	7,167.49	7,525.6	7,902.16	8,297.26	8,712.13
Operating expenses (in $)	3,729.82	3,891.31	3,970.06	4,052.75	4,139.57
Net operating expenses (in $)	3,437.67	3,634.55	3,932.09	4,244.51	4,572.55
Implementation costs (in $)	10,194.00	-	-	-	-
Net income (in $)	6,756.33	3,634.55	3,932.09	4,244.51	4,572.55

reach the expected production, whereas vegetables such as onion and lettuce grow in a faster rate than the expected values. However, due to the fact that the vegetables are for self-consumption, it will not be difficult to provide enough time for the vegetables to reach the corresponding size and obtain better results.

Now we turn to the costbenefit analysis. The results obtained for the projection of the expenses and production over 5 years are presented in Table 3. This table shows evidence of the advantages of using cultivation tables. The macroeconomic effects such as inflation and loss of purchasing power can be minimized.

The projection shows that the 'income' and the 'operating expenses' during 5 years in our system increases slightly. Only in 'Year 1' it is necessary to consider the implementation cost. It is important to realize that this cost can be lower if you include more recycled material. However, just in 'Year 1' the system exhibits economical losses, but for the rest of the years the system presents incomes. As a matter of fact, these incomes also increase slightly.

In comparison, the price of the vegetables in retail markets are subject to several factors as well as costs including the use of pesticides, harvest labour, transportation, refrigeration, distribution and storage that are difficult to predict and might become even higher in highly populated areas far from agricultural centers. In contrast, cost associated with crops from Pop-Up Farms have the potential to be more stable due to their simpler production system.

5 CONCLUDING REMARKS

This work has presented preliminary results from the implementation of a Pop-Up Farm system that consists of two cultivation tables using Mel Bartholomew's method and an ad-hoc irrigation system with rainwater harvesting. It has been shown that it is a relatively easy method in terms of manufacture and operation with the additional advantage of low costs suitable to be implemented in highly populated urban areas.

System production at this point has been estimated using crop heights at early growth stages and comparing it with estimated heights from seed producers. Validation of this methodology will be tested once actual harvest results are obtained and will be published accordingly.

A simple economic model has been used to estimate the potential economic benefits of Pop-Up Farms showing that even with low crop performances the system presents net incomes that increase slightly over time making them a viable economic opportunity. Improvements considered for future designs include the use of a higher proportion of recycled materials that meet the operational requirements of the cultivation tables.

In highly populated areas in developing countries, such as Mexico City, Pop-Up Farms present multiple additional benefits such as reducing the energy usage in crop transportation, refrigeration, distribution and storage; reduce packaging-related waste; and increase poor families access to fresh vegetables contributing to an overall improvement in nutrition, food security and health conditions.

For a megacity such as Mexico City, urban agriculture can contribute to a better nutrition and economy since this model can be available for a poor population to have an improvement in terms of food security and health conditions. Human health through food safety and diversified diet is becoming an urgent issue for governments. Urban farming may not be the solution to solve food scarcity, but it is a way to establish condition for permanent upgradable systems able to adapt to an environment that changes day by day.

ACKNOWLEDGMENTS

The authors gratefully acknowledge the assistance of Rodrigo Olivares Ronces in the cost-benefit analysis and for useful discussion about the economic benefits of our system.

P. Burgos Madrigal thanks Pablo David Hernández Guerrero and Luis M. Moran Mata for funding the construction of the system and research.

This research did not receive any specific grant from funding agencies in the public, commercial or not-for-profit sectors.

REFERENCES

[1] Losada, H., Martínez, H., Vieyra, J., Pealing, R., Zavala R. & Cortés, J., Urban agriculture in the metropolitan zone of Mexico City: changes over time in urban, suburban and peri-urban areas. *Environment and Urbanization,* **10(2)**, 37–54, October 1998. https://doi.org/10.1177/095624789801000214

[2] Losada, H., Bennett, R., Soriano, R., Vieyra, J. & Cortés, J., Urban agriculture in Mexico City: functions provided by the use of space for dairy based livelihoods. *Cities*, **17(6)**, 419–31, 2000. https://doi.org/10.1016/s0264-2751(00)00041-x

[3] Torres Lima, P., Rodríguez, L.M. & García Uriza, B.I., Mexico city: the integration of urban agriculture to contain urban sprawl. *City Case Study Mexico,* pp. 364–90, 21 December 2014.

[4] Secretaría de Desarrollo Rural y Equidad para las Comunidades (SEDEREC), Agricultura sustainable a pequeña escala. ASPE, 2017 [Online]. Available: https://sederec.cdmx.gob.mx/programas/programa/programa-agricultura-sustentable-pequena-escala-de-la-ciudad-de-mexico (accessed 7 June 2018).

[5] Pérez Rodríguez, G.I., Delgado Morales, R. & Bernal Negrete, L.G., Agricultura urbana: una alternativa sustentable para el desarrollo genereal. In *21 Encuentro Nacional sobre Desarrollo Regional en México*, Mérida, 2016.

[6] Instituto Nacional de Estadística y Geografía (INEGI), Information by entity. 2018. [Online]. Available: http://cuentame.inegi.org.mx/monografias/informacion/df/territorio/clima.aspx?tema=me&e=09 (accessed 1 June 2018).

[7] Food and Agriculture Organization, Food Security indicators. Food and Agriculture Organization of the United Nations, October 2017 [Online]. Available: http://fao.org/economic/ess/ess-fs/ess-fadata/en/#.W4Vqt2dBCFM (accessed May 2018).

[8] Bartholomew, M., *Square Meter Gardening*, Barcelona: Blume, 2016.

[9] *ABC del Huerto Urbano* [Interview]. 4 May 2018.

[10] Quiñonez Ferandez, M.V., *Use of Coconut Fiber as a Substrate*, Rafael Landívar University: Guatemala, 2014.

[11] Agricultural Clinic Group, Earthworm humus, vermiculture, 2018 [Online]. Available: http://bioagrotecsa.com.ec/lombricultura/humus-de-lombriz.html (accessed 3 June, 2018).

[12] Rancho los Molinos, Vita, Buscador de Semillas. Rancho los molinos, 2018 [Online]. Available: http://losmolinos.com/vita/vita.html#one (accessed 5 June 2018).

[13] Instituto Nacional de Estadística y Geografía (INEGI), Encuesta Nacional de Ingresos y Gastos de Hogares, 2016 [Online]. Available: http://beta.inegi.org.mx/contenidos/proyectos/enchogares/regulares/enigh/nc/2016/doc/presentacion_resultados_enigh2016.pdf (accessed 2018).

[14] Instituto Nacional de Estadistíca y Geografía (INEGI), Comunicado de prensa. 2016. [Online]. Available: http://inegi.org.mx/saladeprensa/boletines/2017/enigh/enigh_08.pdf, 2016 (accessed 2018).

[15] SNIIM, Sistema Nacional de Información e Integración de Mercados. SNIIM, 17 June 2018 [Online]. Available: http://economia-sniim.gob.mx/2010prueba/FrutasFec.asp (accessed 17 June 2018).

[16] Servicio de Administración Tributaria (SAT), Salario Minimos 2018 [Online]. Available: http://omawww.sat.gob.mx/informacion_fiscal/tablas_indicadores/Paginas/salarios_minimos.aspx (accessed 30 Agosto 2018).

[17] Morrow, J., *Vegetable Gardening for Organic and Biodynamic Growers*, SteinerBooks, Incorporated, 2014.

[18] Barkin, D., La gestión del agua urbana en México. In *La gestión del agua urbana en México: retos, debates y bienestar*, Guadalajara, Universidad de Guadalajara, pp. 3–6, 2006.

[19] NASA Surface meteorology and Solar Energy, Atmospheric Science Data Center, 2018 [Online]. Available: https://eosweb.larc.nasa.gov/cgi-bin/sse/retscreen.cgi?email=rets%40nrcan.gc.ca&step=1&lat=19.507788&lon=-99.097626&submit=Submit (accessed 1 June 2018).

[20] Dieleman, H., Urban agriculture in Mexico City: balancing between ecological, economic, social and symbolic value. *Journal of Cleaner Production,* **163**, pp. S156–S163, 1 October 2017.

[21] Forbes, Forbes México, 19 February 2015 [Online]. Available: https://forbes.com.mx/mexicanos-gastan-50-del-ingreso-familiar-en-transporte/ (accessed 3 May 2018).

[22] Instituto Nacional de Estadística y Geografía (INEGI), Indice de precios. INEGI, 27 May 2018 [Online]. Available: http://beta.inegi.org.mx/app/indicesdeprecios/Estructura.aspx?idEstructura=112000200040&T=%C3%8Dndices%20de%20Precios%20al%20Consumidor&ST=Inflaci%C3%B3n (accessed 17 June 2018).

[23] SEDESOL, Gobierno de Mexico, Secretaria de Desarrollo Social, 2017 [Online]. Available: https://gob.mx/sedesol (accessed Junio 2018).

[24] Secretaría de Hacienda y Crédito Público (SHCP), Servicio de administracion tributaria. SAT, 21 December 2017 [Online]. Available: http://omawww.sat.gob.mx/informacion_fiscal/tablas_indicadores/Paginas/salarios_minimos.aspx (accessed 17 June 2018).

CULTIVATING REFUGE: THE ROLE OF URBAN AGRICULTURE AMONGST REFUGEES AND FORCED MIGRANTS IN THE KURDISTAN REGION OF IRAQ

M. TOMKINS[1], S. YOUSEF[2,3] , A. ADAM-BRADFORD[4], C. PERKINS[5], E. GROSRENAUD[4], M. MCTOUGH[4,5] & A. VILJOEN[1]
[1] School of Architecture and Design, University of Brighton, UK.
[2] Department of Recreation and Ecotourism, College of Agriculture, University of Duhok· Iraq.
[3] Botany and Modelling of Plant Architecture and Vegetation (AMAP), Université de Montpellier, France.
[4] Centre for Agroforestry, Water and Resilience, Coventry University, UK.
[5] Southern Methodist University, Dallas, US, [6] United Nations Development Programme, Iraq.

ABSTRACT

Refugee camps are born out of chaos and crisis, characterised as short-term responses with little in the way of planning for long-term living. However, it is becoming increasingly apparent that within protracted refugee situations, all too often these camps morph into 'accidental cities', where an accelerated everyday urbanism transforms tents into streets lined with self-built homes. Within the camps of northern Iraq, displaced Syrian refugees are finding innovative ways to incorporate urban agriculture and agroforestry into these unintended but now permanent settlements. Largely unsupported and often in conflict with the initial disaster response planning for camps, Urban Agriculture (UA) flourishes at a household level, providing access to fresh food, healing spaces from trauma, and creative place-making practices. Using lessons learnt from three years of practical fieldwork developing and supporting UA in camps located in northern Iraq, this paper demonstrates that with or without institutional support home gardens emerge at every stage of camp development as a vital yet little-discussed and even less planned practice. The paper argues that refugee settlements, home to millions worldwide, need to be seen as both urban and permanent, with home gardening and agriculture as a core response at the point of crisis, or risk developing, by default, into unsustainable – slum-like – cities of the future.

Keywords: Ethnobotany, greening innovation, home gardens, Iraq, Syria, Kurdistan, agroforestry, refugee camps, SuDS, urban agriculture.

1 INTRODUCTION

Worldwide geopolitical conflicts generate mass movements of internally displaced persons (IDPs), refugees and forced migration. In the context of the Syrian civil war, it has created one of the most complex and expansive humanitarian crises – that has displaced many Syrians across the neighbouring countries of Turkey, Jordan, Lebanon, and the Kurdistan region of Iraq (KRI). The landscape of the KRI, where Syrian refugees find themselves, is mountainous and located in a semi-arid zone with harsh environmental conditions (e.g. mean daily temperature from June to August exceeded 40 °C). Tents used in refugee camps exacerbate this heat: inside temperatures exceed 50 °C, making them uninhabitable. This is often compounded by frequent dust storms in the summer and flash flooding in the winter.

A large proportion of Syrians find their way to the camps in the Dohuk region, such as Domiz, which is the largest in KRI [1]. Some have travelled from the Mesopotamian Region of northeast Syria and find themselves in the upper Mesopotamian plains of the Kurdistan Region of northern Iraq. At the initial opening of Domiz Camp, in 2012, the landscape and area looked like a semi-desert with no signs of life, green spaces or trees. Today, thanks to agroforestry and UA it resembles other Mesopotamian towns, with gardens, parks, and all other necessary urban infrastructure.

DOI: 10.2495/DNE-V14-N2-103-118

Due to long-term conflicts becoming more commonplace, it is now accepted that 'refugees are spending longer periods in exile and increasing attention is now being paid to the rise of protracted refugee situations' [2]. A protracted refugee situation (PRS) is defined as a displacement for more than five years from the primary displacement where there is little chance of a long-lasting solution. Syrian refugees largely fit into this definition as the 2011 Syrian civil war enters its 8th year.

1.1 Project overview

Several international Non-Governmental Organisations (NGOs) have recognised the importance of developing agriculture and greening in the context of migration and crisis [3]. For example, in 2000 the United Nations High Commissioner for Refugees (UNHCR) and FOA provided 15,000 families in Tanzania with seedlings [4], and in 2005 UNHCR distributed 200,000 seedlings as part of their Greening Camps Programme [5]. More recently, Save the Children alongside the World Food Program helped to develop gardens in Za'atari Camp, home to more than 70,000 Syrian refugees.

Building on such work, this paper will discuss UA project work in KRI from October 2015 to May 2018, focusing on Domiz Camp, together with supporting fieldwork in two additional refugee camps, namely Domiz 2 and Gawilan, all home to a majority of Syrian refugees. Furthermore, in the latter part of 2017 to May 2018, the team facilitated home garden development in three IDP camps, Kabartu 1 and 2 and Essian, with a majority Yazidi population (Table 1). While the project developed a wide range of practices, including cultivating home gardens, building greenhouses and creating communal gardens, this paper will focus primarily on the aspects of home gardening and tree planting in Domiz, which is situated approximately 70km from both the Syrian and Turkish borders (Fig. 1). Opening in late 2012, with more than 34,000 Syrian refugees, it now contains approximately 29,000 residents (2018).

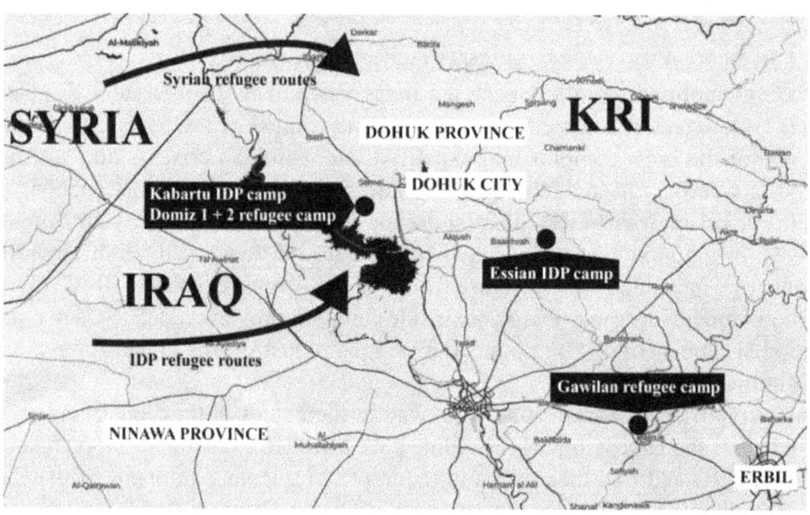

Figure 1: Regional map of Iraq and Syria show refugee and IDP camps.

Table 1: Ethnographic and population data of refugee camps.

Camp name	Ethnicity	Population		Date of profile
		N° families	N° individuals	
Domiz 1	Syrian Kurds (Muslim)	5,721	29,100	August 2018
Domiz 2	Syrian Kurds (Muslim)	1,892	8,734	April 2018
Gawilan	Syrian Kurds (Muslim)	1,950	10,200	2018
Essian	Iraqi Kurds (Yazidi)	2,720	14,497	2018
Kabartu 1 & 2	Iraqi Kurds (Yazidi, Muslim)	2,541	13,511	April 2018

Source: Board of Relief and Humanitarian Affairs, private communication 2018.

- The field team consisted of four researchers, two of which were based in the UK, one based in Iraq, and one based in the US, supported by a US creative director. The work focused on the practical implementation of UA at a household and a communal level and was funded by a small international NGO, known as Lemon Tree Trust. The project was driven by the desire to advocate for and demonstrate the potential for UA as part of a crisis response as well as within permanent settlement development. Furthermore, an additional objective was the building of a sound evidence base for UA in refugee camp settings which could then be up-scaled by other larger national and international NGOs.
- It was also hoped to alter regional or international discourse and policy around UA or broader 'Greening Innovation' practices. The term Greening Innovation was used to encompass everyday spatially innovative practices that 'use environmentally friendly, climate-smart technologies and practices to grow food, plant trees, and produce energy, and to convert our waste into resources using productive closed-loop systems that actually build rather than exploit their natural resource base' [6]. Greening Innovation can be linked to concepts of 'spatial sovereignty', which emphasise the need to have autonomy over space as a precondition for food security or even food sovereignty [7]. This distinction emphasises that food gardens require both the transforming and occupation of space prior to growing, something that disenfranchised and traumatised newly arrived refugees might struggle to embrace [8].

1.2 UA definitions in the context of forced migration

Within the context of the refugee camp, working definitions of UA may take on divergent and more nuanced definitions than those that focus on materials and productivity such as those offered by Bakker [9]. For example, Perez-Vazquez notes UA should not only include just material benefits but also take account of health, recreation, and relaxation [10], emphasising the non-productive aspects of UA outside of commerce, which can include ecological functions such as biodiversity and micro-climate regulation, and cultural aspects such as leisure, cultural practices, and creative place making [11].

Table 2: Urban agriculture classification area type within refugee camps.

Area types	Comments
Open spaces	Any safe unused open space
Camp boundaries	Areas immediately inside camp fences
Households	Home gardens from small container gardens to dense agroforestry plots
Green open spaces	Camps parks, roadside tree planting
Community gardens/farms	Community-based approach with social cohesion and welfare objectives
Tree plantations/orchards/woodlots	Community-based approach with linkages to energy and livelihoods
Flood plains	Includes wetlands which can be integrated with UA
Institutional spaces	Areas around offices, administrative spaces, warehouses
Schools	School gardens and farms; can be integrated into citizen science projects
Communal growing areas	Allotment type gardens that are cultivated individually
Peri-camp spaces	Areas immediately after camp fences (often associated with waterlogging) that may also include host community
Open but controlled spaces	Areas allocated for emergency responses such as cholera isolation areas
Temporary spaces	Areas allocated to future infrastructure but currently empty

UA in the context of refugees in Iraq therefore includes both vegetables and flowers, as well as creative acts such as sculptures and decorations. However, there is no guarantee that refugees will get a space to garden in despite guidelines that call for such spaces. For example, in 2017 UNHCR issued official planning standards where they suggest 'A minimum surface area of 45 sqm per person, which includes 15 sqm allocated for household gardening which should be included in the site plan from the outset' [12]. These standards for spatial provisions for gardening are also recognised within the Sphere Project Guidelines, where there is an endorsement for the provision of 'limited kitchen gardens for individual households' [13]. What is important therefore, is the need to support, create and safeguard the use of open productive space to preserve activities that bring resilience so that communities might be better prepared to absorb, recover or prepare for future eventualities (see Table 2). This is critically important within the northern area of Iraq where conflict, in the form of Islamic State of Iraq and Syria (ISIS), was ever present.

2 PROJECT GOALS

During 2015, the team conducted several months of preliminary fieldwork with four primary goals: (1) Identify existing gardens and gardeners; (2) Recruit a refugee field team;

(3) Understand land distribution and availability and; (4) Ensure that potential interventions do not disrupt local refugee development and businesses. Much of the groundwork was achieved through a simplified form of participant observation (PO) where the team would walk, and spend time, in the camp with minimal intervention. Broadly, PO requires the researcher to participate with people in their natural environment, working 'with' rather than 'on' the local community [14]. Through a process of natural integration in the camp, we were able to identify existing gardens, gain an understanding of regional and cultural gardening practices and gain the confidence of gardeners and families.

Through this trust building process, we were able to recruit future project team members as well as gatekeepers from within the refugee community. One such example was the identification of a small plant nursery, where residents would buy flowers, trees or seeds. It was evident that any distribution of seeds or trees would disrupt this nascent business. We therefore approached the owner, with a proposal for his business to become a UA distribution hub where residents could collect seeds and trees for the development of home gardens. The owner benefitted hugely from the influx of customers evidenced from the growth of his business into a thriving plant nursery. The project benefited because we gained the trust of the community and because the continued distribution of trees and seeds did not depend on the presence of international team members.

2.1 The accidental city and accelerated urbanism

Spatially, Domiz Camp is largely built on a street grid format. However, settlement density is very uneven where some dwellings are crowded together and have no outside spaces, while others have extensive space. The original camp layout grew from tents in rows along streets of varying widths. However, as structures were portable, refugees could move tents or create extensions to a tent within the grids by using empty plots or creating extended family dwellings. By contrast, the dwellings in Gawilan Camp resemble that of rigidly planned housing developments, with almost identical spacing between houses and open space within the walls. While this presented less opportunity for self-development, the garden spaces are reasonable and most households made use of the outside space with some form of planting.

The right of refugees to modify the dwellings stands in contrast to the camps that are home to internally displaced persons (IDPs), where the residents are forbidden to alter or augment housing plots. On the basic spatial point, we often found that IDP camps had fewer examples of UA (home gardens or allotment style gardens) than in refugee camps such as Domiz. IDPs as a rule stayed within the tent plot they were given, compounded by the fact that there was a greater expectation of them returning. As a result, Domiz Camp has rapidly transformed itself from a tent city in 2015 to a self-built informal settlement with no temporary dwellings by 2018. This process has been fully supported by UNHCR, Peace Winds Japan and the local government, but also driven by refugees themselves where they have the economic resources to build an improved structure.

From first-hand experience in Iraq (mid 2018), the conversation between UNHCR and local government is now centred on full integration of camps with host strategic planning, both structural and social, where there is little expectation of Syrian refugees returning to a post-civil war Syria. Specifically regarding UA, this process has been mapped out by VNG international in Jordan, whose report 'Linking Urban Farming and Urban Planning in Times of Crisis' [15] examines the potential for a city region food system in Mafraq, the host town close to Za'atari refugee camps. As refugee camps continue to rapidly integrate with hosts, there are opportunities for innovative planners, architects and designers to contribute and learn from this process in the context of development under crisis (with food at its core) creating necessary feedback to questions that are being asked in the southern global context.

This accelerated or rapid urbanism in Iraqi camps presents both opportunities for UA and also challenges. The tension inherent in transformation is that competition for space will inevitably move away from food production to pressing matters of housing the growing population, largely because the camp itself is no longer in crisis mode. The rapid rate of construction favours market forces where refugees have decided to stay put and find jobs, and where birth rates are high. In this situation, living space is both an expression of income but also at a premium to house the next generation. Living space is finite while food provisions can be externalised to the markets and shops in the camp where fresh food is expensive but readily available.

2.1.1 Interviews with refugees

Data was collected in 2017 in the form of interviews and focus groups for a two-week period in January and again for a 10-day period in May 2017. Interviews were limited ($n = 165$) due to time constraints although were representative of families that were deeply involved with the project, 26 of whom could be considered key informants. Interviews were tape recorded and translated into English from either Kurdish or Arabic and then transcribed. This data was reinforced by the door-to-door work of the team where interviews could be made regarding current gardening practices. The data-collection was approached with the goal of compiling both quantitative and qualitative information from refugees living inside Domiz Camp. Data collection was carried out using a mixed methodology composed of four tools: (1) Ground canvassing to assess the current state of urban agriculture and gardening inside Domiz Camp; (2) Qualitative focus-group discussions (FGDs) divided by gender ($n = 2$); (3) Key informant interviews with families ($n = 10$), refugee participants from the 2017 garden competition ($n = 16$); (4) Survey data collected from all 2017 participants concerning what their gardens contained and whether they had gardened previously ($n = 139$).

The interviews used open-ended questioning to ask about what gardening means for people, as well as specific questions to determine if people were new gardeners or had a garden back in Syria. We also asked, 'Is there anything else you would like to say about gardening or agriculture?' to allow for more open discussion about how people felt about gardening in refugee camps. Key themes that emerged from the interviews were fairly common to the practice of gardening, such as the therapeutic value of gardening, the use of space for health and privacy and community. People also used gardening as a release from frustrations or boredom.

For example, one male respondent (aged 53), states, 'Gardens bring peace of mind: old men and women have a place to sit and talk, the war brought us many things and we need to remember and to spend time together talking it over'. A female respondent states, 'I live with a lot of pressure here, my daughter is divorced, my husband is sick, I grow my garden and it makes me feel better.' Gardens also provided an outlet for meaningful activity, as many in the camp dealt with a lack of employment opportunities, as one male (aged 34) states, 'Jobs are important, but cultivation is too, seeds and trees are better than free vegetables, give us seeds and trees.' Additionally, gardens often provided a reminder of home, 'This garden reminds me of my childhood, my land, it also benefits me for food, essentially it connects me to my homeland,' said a female respondent, aged 48. Perhaps most significantly, people spoke of the need to make space and to create something beautiful within the confines of the camp, where the garden becomes a microcosm of the larger potentials. For example, 'Gardening is a home thing, a chance to create my own place,' said one male respondent, aged 45. Through these interviews, the team were able to achieve greater empathy with refugees and their aims with their gardening practice. This understanding helped to balance the earlier discussion of definitions of UA, where there is less emphasis on the 'production functions' of UA, and encompass some of the more nuanced cultural functions such as recreation, cultural heritage, and place-making.

2.1.2 Summary of project outcomes

During the period October 2015 to May 2018, in the Kurdistan Region of Iraq, the following summary of outcomes will focus primarily on Domiz Camp, while also including less intensive fieldwork performed in additional refugee camps, namely Domiz 2 and Gawilan, and three IDP camps, Kabartu 1 and 2 and Essian. While the project developed a wide range of practices, including cultivating home gardens, building greenhouses and creating communal gardens, the outcomes are largely centered on aspects relating to the monitoring, facilitation and support of home gardening and tree planting in Domiz, along with the corresponding environmental and psychosocial impacts of these efforts. While quantitative evaluation of the project was achieved by a more straightforward analysis of the number of trees distributed, planted and the increase in actual garden space, the social impacts were largely evaluated by an analysis of ethnographic interview data, as well as detailed observations of community interactions revolving around new spaces of cultivation.

3 HOME GARDENS, TREES AND PLANTS

While a wide range of practices were developed, such as greenhouses and community demonstration gardens, the discussion will focus on the home garden development, as this was a key practice and one that brought the team in day-to-day contact with families. It also necessitated an interaction with the creation of 'on street' spaces, together with water management, and tree planting. The creation of gardens that face the street immediately improve social interaction and can create a cascade within neighbourhoods as adjacent families follow suit. Moreover, potential home garden spaces or spaces around or close to tents are reasonably within peoples' everyday capacity to create, and require less permissions from the camp authority. From experience, larger practices such as communal or market gardens require people to seek permission, raise cash and negotiate with camp or local authorities and are therefore less likely to happen.

The majority of camp residents are Syrian Kurdish and historically Syrians have a deep connection to creating gardens – an attachment that seemed to be especially amplified in the precarious and desolate setting of a refugee camp. This inherent cultural and religious attachment to gardening emerged as a way to exercise some control of immediate surroundings at a time when control over broader events has been taken away, as well as a reflection of traditional belief systems. It was not uncommon while working in the camp, to meet refugees who had managed to bring seeds with them when fleeing Syria or had later managed to return to Syria and sourced seeds during these trips. Local seeds available were generally imported and generic to northern climates, with European brands such as Franchi, dominating.

3.1 The invisibility of garden spaces

While home garden space forms part of guidelines for camp design, its implementation is uneven, unenforced and in some cases, non-existent. This is compounded by the lack of spatial data in the form of maps. For example, initially, UNHCR might produce a map to support camp layout, along with intermittent updates as the camp develops. Such maps show significant structures such as large buildings, schools, medical centres, bore wells and roads. However, the scale of such maps does not record the spaces around tents (and later permanent dwellings) that might support household food production. The invisibility of these spaces from official documentation adds to their precarious nature. For example, during surveys in regional camps, it was not uncommon to hear camp management or staff comment that 'no one gardens in the camp', despite gardens being in evidence in the camp.

This invisibility is at both the level of everyday life but also at an institutional level. Regarding the former, households might make a decision to initially have a garden to create a sense of home or for a place to relax, provide shade or for basic food production. Later this garden space will be subsumed as the need for extra rooms for expanding families becomes a priority. The appearance and disappearance of these domestic, fragmented and intimate spaces remains unrecorded. Regarding the latter, camp management may disregard the importance of everyday household food production and gardening as a practice not only because it remains largely invisible, but because they are tasked with providing the larger pieces of infrastructure, and maintaining a flow of goods and services in the camp in relation to UNHCR and NGO funding. Thus, the everyday practice of 'cultivating refuge', whereby refugees through spatial innovation create much of the public realm, and street design goes unacknowledged and recorded.

The project work pursued during the period from 2015 to 2018 therefore sought to contribute to the process of building and cultivating already established by refugees rather than to bring in design professionals or impose constraints around aesthetics. The team focused on providing basic tools, seeds, and trees to refugees to supplement and encourage the home gardens already underway or provide a level of advocacy. Despite northern Iraq having a volatile economy, the team found it reasonably easy to access gardening supplies in local agricultural shops. Hand tools were variable in quality but essential given the lack of resources available to refugees and the poor soil quality within the arid landscape of northern Iraq.

3.2 Categories of home gardens

Water features were common because they provided a focal point in the garden and essential cooling in the hot summer. The fountains are often constructed from found materials left over from camp construction such as pipes and tiles and used small pumps to recirculate water. Infrastructure created to support gardening used concrete to a large extent to create beds, water features of sculpted walls with patterns or images. These could be extensive, covering an entire face of a single storey house or garden and incorporated water features or raised beds for flowers or food. Decoratively painted, such homes create a contrasting street scene of differing patterns and development furthering the act of creativity and autonomy in camp place-making.

In all, the team recorded 16 categories of garden spaces, although these are not mutually exclusive as families may incorporate different practices within one space. Also, practice may be dependent on size where some families had extensive space for incorporating different aspects and others responded to spatial constraints by limiting themselves to one category of garden (Table 3).

Some of these categories are self-evident such as trees or a rose garden where gardens feature several trees used for shade or shelter. Roses are particularly prized by Syrians and several households chose to grow a single rose bush, which would be tended and pruned to produce a show of flowers. Beyond the ornamental, the potential economy for rose petals for

Table 3: Sixteen categories of home garden types recorded in northern Iraq.

Home garden type	Main characteristics	Main benefits
Trees	Tree planting around dwelling or plot.	Increase shade with light weight dwellings, provide shelter from elements, privacy, food, amenity.
Innovation	Imaginative transformation of often limited materials, space and plants.	Creative and therapeutic practice providing a greater sense of home
Biodiversity	Use of local and companion plants to maximise biodiversity without necessarily providing an aesthetic dimension.	Benefits are often external to the dwelling, where such gardens might become educational or aid local biodiversity.
Single planting (e.g. Rose)	Exclusive cultivation of a significant plant (e.g. roses or calendula).	Reconnecting with and symbolic of home or cultural practice such as a ceremony or medicinal requirement.
Container	Gardening on hardstanding.	Using containers (often recycled) to grow when no soil bed is available.
Vegetable	Exclusive cultivation of edible plants.	Provide nutrition, bridge food security and food sovereignty issues, potential income.
Ornamental (mixed planting)	Ornamental planting without food.	Creating a leisure space which may remind people of home or create a new sense of place.
Decorative	Decorative garden might contain structures and decorative elements without the use of plants.	Creates amenity space that may have no immediate food output but creates spatial potential in future.
Recycling	Use of found or scavenged materials.	Helps reduce waste in camps and also demonstrates an affordable use of materials where resources are scares. This is also linked to innovation garden above.
Neighbourhood	Linked gardens between dwellings or tents that provide a visual continuum or shared resource.	Improved public realm, often create feeling of overall design initiative, and help with shared resources and biodiversity if plants.
Limited space	The use of vertical or walled space, incorporated in dwelling structure for example.	Maximise small use of space.

(Continued)

Table 3: Sixteen categories of home garden types recorded in northern Iraq. *(Continued)*

Water feature	Use of fountains or integrated water flows.	Cooling in summer with emphasis on leisure and healing. Potential for recirculation systems such as aquaponics or hydroponics.
Ecological	Use of greywater or recycled matter such as compost.	Contributes to environmental sanitation.
Intensive growing	Planting a high diversity of trees, shrubs, and ground plants within a small area.	Improved resilience through broad range of plants. Potential to act as a local resource site for other households.
Cash crop	Single cultivated crop for home consumption or market place.	Production of crops on large scale barter or cash.
Street garden	Use of sites outside but close to home.	Externalise production when the dwelling does not provide enough space. Provide a public statement to help inspire others.

the production of either rose oil or rose water was felt to be undervalued. Rose oil is a valuable cash crop which is made from rose petals. However, extensive planting, tending, and watering are required to produce the volume of petals for commercial scale production. One solution, which is currently being pursued in Domiz Camp, is to distribute thousands of damask rose bushes at a household level, creating a type of atomised farming within and across camps with centralised harvest and distillation.

3.3 Details of home garden categories

These kinds of solutions aim to bridge the divide between household practices and potential livelihood strategies, whereby the everyday tactics of domestic gardening can connect within broader strategic interventions (Fig. 2). Tree planting presents another example, whereby single trees planted at a household level for shade or fruit begin to aggregate into wider agroforestry solutions. These have the potential to remediate poor water drainage, improve the overall streetscape and reduce the need for extensive cooling in summer due to the lightweight nature of refugee homes. Neighbourhood gardens are another example of a bridging practice. These are categorised as gardens where families have deliberately created continuous and connected spaces through social cascading. These are sometimes in harmony with neighbours where similar materials and styles are adopted, are more fragmented, where each family has embraced a distinct style and planting system. Where these are on public show, they create a streetscape which immediately transports one out of the confines of the camp environment, and would not look out of place in less impoverished suburban cities.

Gardening in limited spaces was very common where residents might only have a metre-wide strip or concrete walkway to grow on. Such spaces are closely linked to the categories of innovation and container growing. One such example shows a front garden measuring one

Figure 2: Decorative and productive food garden, Domiz Camp (copyright Mikey Tomkins 2016).

metre by five metres (Fig. 3, left). In this space, the family has used recycled guttering to create a vertical garden to grow onions and garlic. On the floor it grows salad and herb crops, while the front section, decorated by recycled wooden crates, has decorative flowers, shaded on the roof by more flowers and vegetables. Figure 3 right shows a concrete walkway edged with container grown plants and trees, which aim to provide decoration and some privacy which is hard to achieve in the camp due to the lightweight nature of the buildings. Also, evident here is the juxtaposition of a wall made from UNHCR branded tarpaulin that creates a curtain with a wall of the neighbouring house, which is built to a higher standard of development.

Food gardening was widely evident but not dominant in camps. Food production ranged from one family that was growing a single garlic crop for cash in a limited space, to micro-allotment gardens of multiple vegetables, or leafy vegetates interspersed with ornamentals. While displaced people suffer from endemic poverty, refugees in particular, who may not be able to work or have bank accounts, food, both fresh and processed, is widely available yet restricted by cost and uneven distribution. Food growing therefore, while it might be critical

Figure 3: Left, one metre by five metres garden. Right, concrete container garden (copyright Mikey Tomkins 2016).

to some families, is generally supplementary rather than primary. This was evidenced by the intercropping growing of herbs, salad, or alliums rather than staple crops or long seasonal field crops. Within the category of food gardens, we should also include livestock, which was largely represented by rabbits and chickens but again this is unlikely to be primary and would largely provide occasional meals.

Street gardens describe families who have created garden spaces outside their dwellings and occupied a patch of ground close to home, often with neighbours. These gardens were not evident during the 2015–2016 season but began to emerge from 2017 onwards. This may be indicative of families settling in to camp life and making more long-term investments in gardens. Street gardens are more precarious because camp management can remove them without notice if they conflict with broader strategic planning for infrastructure. One garden the team visited in the category of intensive growing demonstrated the capacity of a small space to contain multiple planting schemes ranging from a canopy, to shrubs, to ground level growing. This small plot measuring five metres by 20 metres, contained 54 species of plants (Table 4), including 16 trees, one of which was a banana plant which was quite an innovation for the region.

An important aspect of the project was the need to work with the community rather than on the community. Some tree planting was already evident in the camp, the work of an early inhabitant of Domiz, Sami Youssef. Sami is a refugee but also a lecturer-researcher at the Faculty of Agriculture at the local Duhok University. His personal initiative sought to encourage other refugees and IDPs to plant fast growing shade trees that can symbolise home and encourage place-making within 'ordinary' life, representing a future that many refugees have lost through the war. The main objective was to improve the urban greening urban inside the camp via creating shaded spaces surrounding tents and thus creating a more clement urban microclimate. More than 2,000 fast-growing trees were bought from public nurseries (from University of Duhok and Directory of Agriculture) then distributed inside the Domiz Camp.

Unfortunately, diverse voices were reluctant to encourage the initiative. Refugees and local authorities alike viewed tree planting as a sign of permanence – saying for example that '*the camp is a temporary stage in our life and here is not our home, we will go back to our home in a very short time. Why should we plant trees here...*'. Moreover, the lack of understanding of the functional roles of the trees in urbanised areas has created inertia, partly explained by the limitation of space and water sources inside the camp. The continuation of this project by Sami Youssef, complemented by the implementation of projects discussed in this paper, alongside the French Red Cross and Mercy Hands, has made Domiz a regional forerunner in the use of UA and greening innovation. While Domiz Camp is out of crisis, refugees are also recognising that the Syrian conflict will not be resolved in the near future. Consequently, they are investing in the construction of their home and gardens, planting olives, mulberry, lemon, fig, grapes, and some medicinal plants such as mallow, mint, balm, marshmallow, and rosa, together with flowering ornamental plants such as roses, and jasmines.

3.4 The role of agroforestry

According to the Food and Agriculture Organization of the United Nations (UNFAO), 1.5 billion people worldwide benefit from trees in a direct or indirect manner [16]. Directly, trees are an important source of food through the supply of nuts and fruits. Indirectly, they supply various materials such as fuel wood, timber, oils, resins, tannins, pigments, latex, fibres, wax, honey, medicine, pesticides, and fodder. Incomes generated can make a significant contribution for households that are food insecure because of low employment opportunities. Despite their

Table 4: Home garden with 54 trees, vegetables, herbs, ornamental and wild plants.

Trees (n = 16) (common names)	Herb/ vegetable (n = 17) (common names)	Ornamental (n = 18) (common names)	Wild (n = 3) (Latin names)
Grape	Lettuce	Dog rose	Bermuda grass (*Cynodon dactylon*)
Mulberry (white)	Basil (purple)	Damask rose	Medick or Burclover (*Medicago sp.*)
Mulberry (black)	Basil (green)	Rose sp.	Sow thistle (*Sonchus olearaceus*)
Olive	Mint (wild)	Marigold (purple)	
Fig	Mint (cultivated)	Marigold (yellow)	
Banana	Rocket salad	*Calendula* (yellow)	
Chinaberry and/or Umbrella tree	Garlic	Dawedia	
Prune	Onion	marvel of Peru (*Mirabilis jalapa*)	
Apple	Parsley	Lily (white)	
Orange	Green pepper	Lily (red)	
Lemon	Green paper (long)	Lily (yellow)	
Pomegranate	Chili	Sunflower	
Fern	Aubergine	Diantus sp.1	
Palm nut	Cucumber	Diantus sp.2	
Peach	Maize	Honeysuckles (*Lonicera*)	
Ornamental tree[1]	Tomato	Ornamental shrub[1]	
	Camomile	Chrysanthemums or Chrysanths	
[1] Unidentified		Ivy	

various benefits, such as fodder and shade for livestock, trees are not always included when designing interventions in camps that aim to contribute to food security and livelihoods. Trees are often overlooked because they do not provide instant relief when compared with food aid. However, while food aid assists in the short-term it generates dependency and hinders long-term host development and solutions. The benefits of planting trees should therefore be considered as a long-term strategy for both the immediate refugee beneficiaries, and the local host communities. The work of Sami Youssef above, clearly demonstrates this. Sami's expertise in understanding the regional plant species means that the correct type of tree is planted. Whatever might happen to Syrian refugees in the future, the host communities can be certain to inherit several thousand trees.

3.5 The role of sustainable drainage systems (SuDS)

Further related developments include the implementation of SuDS in Gawilan Camp (Fig. 1). SuDS is designed to convey water from its source, such as domestic greywater and/or rainfall, through a number of 'devices', which at each stage contributes to control and reduction in flow rates, while improving the water quality through pollution reduction, and in the case of Gawilan Camp, also reducing erosion as surface runoff is reduced. SuDS are important because all homes in camps produce greywater and refugees are allowed to use this water directly on plants. Conversely, they are not permitted to use greywater once it joins a communal flow in open street drainage channels that sometimes forming large waterlogged areas.

Devices incorporated in SuDS can aid community-based urban agriculture. These include initial filtration stages, with 'oil traps' and 'sand filters', which then feed the flowing wastewater into sub-surface aggregate-filled 'trickle trenches', which continue with levels of treatment, while also irrigating 'tree pits' located along the trenches as off-shoots. At the end of the 'trickle trenches', any remaining water then enters the main 'swale', which conveys the treated water into a retention pond which adds to the ascetics and biodiversity of the system. The community played an active role in the site design and a Syrian farmer has been recruited to maintain the urban agriculture and amenity component of the project. The example of the development of SuDs in Gawilan Camp is a clear way in which UA concepts can interlink with and influence wider structural developments, creating a mutual discourse that supports both improvement within refugees' everyday lives and institutional responsibilities for long-term development.

3.6 The role of wild edible plants

WEP ensure food and livelihood security for countless vulnerable families worldwide. Moreover, wild edible diets reflect the regional identity of local communities, and their traditional ecological knowledge like in the Zagros Region [17]. The majority of refugees in Domiz Camp come from the Mesopotamian Region and they combine their experience in agriculture and cultivation with WEP. Refugees (principally women) have the knowledge to still harvest wild edible plants from the Mesopotamian steppic plains surrounding Domiz Camp. The elders also transmit this remarkable ancestral ethnobotanical knowledge of plant nutrition to the young members.

In this steppic grassland habitat, Sami Youssef has catalogued 40 wild edible plants that are used as sources of foods by Syrian refugees which they add to traditional recipes and dishes. These plant species, which are commonly collected by the refugees, are: *Alcea kurdisca, Allium* sp., *Anchusa* sp., *Centaurea* sp., *Crocus* sp., *Eminium spiculatum, Echium* sp., *Geranium tuberosum, Gundelia tournefortii, Malva* sp., *Silybum marianum, Sinapias arvensis, Tragopogon* sp. In the context of forced displacement, the ethnobotanical practices should be considered as an important issue related to enhancing food security and places of hope and dignity.

4 SUMMARY CONCLUSION

Urban agriculture and greening innovation is a powerful force that can help refugees and forced migrants to take control over their lives and local environments. The creation of home garden spaces and UA practices can contribute positively to the architectural process of rapid urbanism in refugee camps, which should be integrated into top-down strategic development. Such integration will be limited if local authorities lack the spatial data and community involvement in preserving smaller fragmented areas that are vital for local domestic and semi-economic UA. Upscaling is also vital, where larger plots suitable for intensive agriculture should be designated for UA and not lost to general development. Moreover, agriculture needs to be considered a vital crisis response strategy throughout the humanitarian and development pipeline, from immediate response to local social cohesion and integration. Building a sound evidence base for UA is vital as we enter a stage of rapid development where professionals and authorities are able to understand how everyday practices and strategic planning shape each other rather than the latter dominating.

Home gardens have also been shown to contribute positively to social and cultural recovery, functioning to preserve memories, knowledge and create sensory interactions vital for trauma recovery within communities. The practice of creating and inhabiting home gardens represents an important link to the past in Syria for many refugees, creating a sense of remembrance for a home and country that many will not return to, as well to as their potential future by creating a sense of belonging and dignity to their new community and home in Iraq.

REFERENCES

[1] Report: Singh, N., Guiu, R. & Higel, L., (eds), *Pathways to Resilience for Syrian Refugees*, United Nations Development programme in Iraq & Middle East Research Institute, 2016.

[2] Book: Betts, A., Loescher, G. & Milner., (eds), *UNHCR: The Politics and Practice of Refugee Protection*, Global Institutions, 2011.

[3] Journal article: Millican, J., Perkins, C. & Adam-Bradford, A., Gardening in displacement: the benefits of cultivating in crisis. *Journal of Refugee Studies*, **31(2)**, 2018. https://doi.org/10.1093/jrs/fey033

[4] Online source: Planting hopes in refugee camps in Tanzania; Food and Agriculture Organisation of the United Nations (FAO) Online, http://fao.org/english/newsroom/highlights/2000/001105-e.htm (accessed May 2018).

[5] http://unhcr.org/uk/news/latest/2005/6/42a06e0d6/greening-refugee-camps-world-environment-day.html]

[6] Online source: Adam-Bradford, A., Tomkins, M., Perkins, C., Hunt, S.S. & Millican, J., Transforming land transforming lives, greening innovation and urban agriculture in the context of forced displacement. Lemon Tree Trust. Online, https://lemontreetrust.org/publications/

[7] Online source: Tomkins, M., Making space for food: everyday food gardening and its contribution to urban agriculture. Online, http://eprints.brighton.ac.uk/12919/1/Mikey%20Tomkins.pdf (accessed May 2018)

[8] Online journal: Adam-Bradford, A. & Osman, M., Tsunami aftermath: development of an indigenous home garden in banda aceh. *Urban Agriculture Magazine* 21, pp. 29–30.

[9] Book: Bakker, N., Dubbeling, M., Gündel, S., Sabel-koschella, U. & De Zeeuw, H.O., (eds), Growing cities, growing food: urban agriculture on the policy agenda; a reader on urban agriculture, 2000.

[10] PhD thesis: Perez-Vazquez, A., The role of allotments in food production as a component of urban agriculture in england. Imperial College Wye, University of London. 2002.

[11] Journal article: Multifunctional urban agriculture for sustainable land use planning in the united states. *Sustainability*, **2**, pp. 2499–2522.

[12] Online source: Camp planning standards (Planning settlements). https://emergency.unhcr.org/entry/45582/camp-planning-standards-planned-settlements (accessed September 2018).

[13] Online journal: Adam-Bradford, A., The sphere project guidelines. *Urban Agriculture Magazine*, **21**, p. 31, 2009.

[14] Book: Spradley, J.P., *Participant Observation*, Holt, Rinehart and Winston: New York, London, 1980.

[15] Online report: Linking urban farming and urban planning in times of crisis. https://ruaf.org/sites/default/files/Mission%20Report%20LogoRep%20RUAF%20SpCiti%20Jul%202016%20FINAL-low%20res%281%29.pdf (accessed September 2018).

[16] Online report: Towards stronger family farms. http://fao.org/3/a-i4171e.pdf (accessed September 2018).

[17] Journal article: Youssef, S., Mahmood, A., Hussein, W. & amp; Véla, E. Montagnes du Zagros, un paradis terrestre aux pratiques ethnobotaniques vivantes. *La Garance voyageuse*, **120**, pp. 41–45, 2017.

METHODOLOGY TO IDENTIFY AND ASSESS AGROECOLOGICAL PRACTICES IN METROPOLITAN AREAS. CASE STUDY, CONCEPCIÓN, CHILE

M. DELPINO-CHAMY, M. ALARCON, S. FERNÁNDEZ & J. SOTO
University of Concepción, Chile.

ABSTRACT

Urban agriculture (UA) is being constantly reviewed because of its significant contribution to urban and metropolitan sustainability (MS). Within the approaches associated with UA, agroecology stands out as a practice that places value on collaborative social networks. Emerging from grassroots organizations that focus on ecological production and food sovereignty, agroecology strengthens short production-consumption chains, therefore increasing the resilience of metropolitan systems, which are under constant threat by natural hazards. The focus of this research is to develop and apply a methodology that identifies underlying agroecological practices, describing their location, state of development, and potential contribution to sustainability within metropolitan areas. A theoretical framework is developed to distinguish agroecological practices from commonly practiced UA; subsequently a proposed methodology is developed to identify and assess these activities based on the following criteria: forms of production, sociopolitical organization, and MS. This methodological approach is used to analyze agroecological practices present in the Metropolitan Area of Concepción, Chile's second most important city, located over a territory with high presence of marshes, riverbeds, and wetlands. Conclusions identify as main strengths of agroecological practices their location on rururban interstices, together with their small scale of production; both aspects contribute to improving MS. Regarding weaknesses, the lack of appropriate planning policies and regulations threaten agroecological practices to disappear under the pressures exerted by contemporary urban development. Finally, particular attention should be given to associative organizations, which have proven to enhance sustainable agroecological outcomes, by increasing employment generation, environmental preservation and local resilience.

Keywords: Agroecology, metropolitan areas, metropolitan sustainability, rururban interstices, sociopolitical organization

1 INTRODUCTION

Metropolitan areas correspond to complex urban systems which are constituted through a set of closely related urban cores, forming a single multi-functional unit [1]. Their administrative structure, composed of different municipal entities; its gravitational weight as a result of the endogenous demographic growth; and the territorial expansion due to internal forces, are factors of complexity that increasingly stress the urban–rural relation in them, and struggle for revisions of the urban development model in pursuit of its sustainability [2]. In Latin America, the metropolitan areas concentrate 57% of the total population, and in the Chilean case there are 10 metropolitan centers recognized by territorial planning instruments (TPIs), such as Metropolitan Planning Schemes or Intercommunal Planning Schemes.

In many of these metropolitan areas, it is possible to observe the coexistence between purely urban activities and others of rural nature, such as agriculture. In most cases, this corresponds to a non-industrialized or medium-scale agriculture.

Various sectors around medium- or large-size cities are used for the development of agriculture, configuring rururban landscapes in the metropolitan borders, or in the interstitial spaces opened in the middle of the urbanized tissue [3]. These areas, in which small-scale agricultural practices are developed, coexist within few meters with urbanizations of different

DOI: 10.2495/DNE-V14-N2-119-130

socioeconomic strata, road, or railway infrastructures, or industrial areas, configuring new forms of metropolitan periphery.

Due to the progress of urbanization over rural areas, associated to the economic growth of nations [4], cities, in general, and metropolitan areas, in particular, demand more and more agricultural products from distant territories. Therefore, some authors [5,6] who define contemporary cities as socioecological systems (SESs) insist in the relevance of linking the social system with adjacent ecosystem resources in order to ensure their sustainability and resilience. However, several models of contemporary urban growth limit the self-subsistence capacity of SES, distancing consumers from food production centers. This makes cities vulnerable to shortages caused by natural or anthropic disasters, limiting their stability [7].

Consequently, the potential of rururban areas, still immersed in metropolitan urban tissues, offers enormous possibilities for the development of agriculture and strengthening of SES.

For the aforementioned reasons, this study is primarily focused on identifying and valuing agroecological practices developing in metropolitan rururban areas. Understanding sustainability is inherent to agroecology, through a constructivist approach [8].

On this basis, the article presents a methodological tool to assess the agroecological components of agricultural practices in metropolitan systems. Later it applies this tool and offers conclusions regarding the assessment of three representative cases studies in Concepción Metropolitan Area (CMA), discussing the strengths and weaknesses of these practices for the sustainability of the metropolitan SES.

2 METROPOLITAN GROWTH AND GENERATION OF AGRICULTURAL PRACTICES IN RURURBAN TERRITORIES

Contemporary metropolization processes are configured through several trends in urban development. On the one hand, a trend that promotes the suburban growth perpetuates the extension of the city over rural territories. These low-density real estate operations end-up configuring an urban system disperse and fragmented in the urban periphery [9]. On the other hand, the trend revitalizing sectors on the consolidated urban area is mainly based on operations which promote vitality and mixture of use through improvements on public space and high-density real estate operations [1].

Between these two tendencies – the low density suburban and the high density hyper-urban paradigms – it is possible to identify the existence of interstitial spaces that have remained halfway between one trend and the other. In the case of the metropolitan areas, "these act as a nexus, on occasions very blurred, between the compact city and the strictly rural territories located at greater distances," as Segrelles declares [4].

Complementary to these dynamics of growth, the topographic conditions of the territory are added. The presence of ravines and water bodies such as rivers, streams, or wetlands determines edge spaces, incapable of sustaining formal urban tissue, despite being often located in strategic areas within consolidated urban centers.

As a consequence of these phenomena of growth and metropolitan urban expansion over a certain geographical territory, a weakening and reduction of the peri-UA practices is observed in the old agricultural areas that surround the city. This results in the emergence of peri-UA practices (in the borders and periphery of the city) or interurban practices (in the interstitial spaces within the consolidated urban tissue) [3].

However, in the metropolitan urbanization processes, these territories both peri-urban and interstitial, suffer the most negative impact of the pressure of urban expansion. On the one hand, uses that fragment the old productive agricultural space that historically surrounded the cities are fostered [10]. On the other hand, pressures generated from real estate development, informal settlements, or the high price of the urban land, causes tensions and conflicts that end-up marginalizing the low-scale agricultural practices which are developed in these rururban areas [4,11].

Consequently, the observation of the phenomenon of areas of rural vocation arises. These have been surrounded and absorbed by the processes of metropolitan urban growth, evidencing how most urban plans and regulatory schemes have been unable to prevent the conversion of rural lands into urban use [12].

Despite these regulatory limitations, UA has proven to offer great contributions to the sustainability of metropolitan urban systems. Proximity between agro-productive areas and consumers, involves a reduction of energy associated to distribution in transportation system, and mobility of citizens towards their supplying centers [13]. Complementarily, non-industrialized agricultural activity contributes with the greening of the city, generating microclimate and improving urban biodiversity [5,14], and local resilience [6,7].

Henceforth, reviewing agricultural practices in rururban areas, with emphasis on agroecological principles and food sovereignty criteria, arise as a strategic approach to improve metropolitan sustainability (MS).

3 AGROECOLOGY AND FOOD SOVEREIGNTY AS STRATEGIC APPROACHES FOR THE STRENGTHENING OF URBAN SUSTAINABILITY

Agroecology started to be studied in the 1970s as a response to the environmental, sociocultural, and economic crisis of the rural areas, as a consequence of the spread of industrial agriculture. It is associated with the concept of ecological agriculture, which is referred to as the application of ecological principles in order to develop sustainable agricultural ecosystems, with a minimal dependence on agrochemicals and external energy subsidies. Agroecology focused its analysis on agro-ecosystems and agro-food systems [15], promoting forms of collective and participative social actions, which encourage ecological food production and local commercialization strategies [16].

In the same manner, the research understands that sustainability is inherent to agroecology, through a constructivist approach [8]. This, unlike the conventional approach, not only includes new paradigms of economic, social, and environmental understanding, but also recognizes the need of local communities to discuss and decide on what they want to sustain, how and by whom.

Gomez et al. [8] and Gutiérrez et al. [17] mention four key conditions to consider agro-ecosystems sustainable: (i) they meet productive goals without compromising the organization of the systems in which they are supported; (ii) they do not depend on inputs foreign to their immediate surroundings; (iii) they are resilient and adaptable; and (iv) they can maintain social organization over time through equitable relations.

This vision is consistent with the principles stated in the Framework for Assessing Natural Resources Management Systems incorporating Sustainable Indicators (MESMIS), which in turn support the approaches of various researchers, such as Gutiérrez et al. [17] and González et al. [18]. MESMIS exhibits the following attributes for a sustainable agroecological practice: (i) productivity, (ii) stability, (iii) resilience, (iv) social equity, and (v) autonomy and cultural adaptability.

All these aforementioned criteria consider technical-productive stability and equitable social relations as key conditions of agroecological practices. Therefore, they strengthen the civil society as a leading agent of the productive and economic processes, for the sustainable organization of cities at both social and ecological scale [13].

Complementarily, special attention should be given to the concept of resilience, given that some authors, such as Álvarez-Salas et al. [19], Folke [5], and Collier et al. [6], have established the study of socioecological resilience, as an alternative to reach agroecosystems sustainability.

Finally, food sovereignty plays a key role in the theoretical approach of agroecology over the sustainability of metropolitan systems. Its conception arises from the farmers and base organizations articulated around the international movement *La Vía Campesina*, in 1993. And a decade later, in the International Forum on Food Sovereignty it was agreed to conceptualize food sovereignty as "the right of people to healthy and culturally appropriate food produced through ecologically and sustainable methods, and their right to define their own food and agriculture systems" [20].

The five pillars of food sovereignty, as defined by Ortega-Cerdà and Rivera-Ferre [21], in coherence with the Declaración de Nyéléni [20], consider: (i) access to resources, (ii) productive model, (iii) transformation and commercialization, (iv) food consumption and right to food, and (v) agricultural policies and organization of the civil society.

Thus, UA designed from an agroecological perspective and aimed to promote urban resilience and food sovereignty constitutes a novel element in the processes of reconstructing urban spaces, with significant sociocultural, environmental and urbanistic benefits [13].

4 METHODOLOGY FOR THE EVALUATION OF AGROECOLOGICAL PRACTICES IN METROPOLITAN AREAS

The proposed methodology is based on the dimensions provided by Soler and Rivera [13], for the design of agroecological indicators. Under this condition, the research analysis and cross methodological criteria defined by the following authors, in order to build an integrative assessment tool for agroecological practices in metropolitan areas: agroecological sustainability as defined by MESMIS; Gómez et al. [8] and Gutiérrez et al. [17] on their conditions to establish agro-ecosystem's sustainability; Ortega-Cerdà and Rivera-Ferre [21] with respect to food sovereignty; and the metropolitan considerations regarding resilience and urban sustainability, as defined by Navarro [3], Folke [5], and Collier et al. [6].

As a result, this study classifies the contributions of agroecological practices to the sustainability of metropolitan systems, under the following dimensions, criteria, and indicators:

4.1 Category 1: Technical-Productive dimension

Agroecology proposes a dialogue of acquaintance between the knowledge of farmers (entomology, botany, soils, and agronomy) and the scientific knowledge (analysis, evaluation, and sustainable management of agro-ecosystems) [16]. As a result of its application, an agroecological technical-productive approach decreases industrial inputs derived from oil, reduces long-distance transportation of foods and promotes the recovery and defense of local biodiversity [13]. Therefore, the assessment criteria defined for this dimension are:

4.1.1 *Criteria associated with FP*

FP. (1) Obtainment of water from natural sources: corresponds to the main natural resource for the development of agroecology. By guaranteeing that water can be reached in a close manner, it is possible to reduce costs of its acquirement and improve technology for its use in the entire agroecological process.

FP. (2) Small-scale production: corresponds to the volume of production, considering three scales: seasonal intensive production; sporadic production which is carried out in some seasons of the year; and family self-consumption, which does not consider commercialization.

FP. (3) Low contamination and use of innocuous inputs: this is referred to both use of inputs and waste generated by the production, which do not contaminate basic resources such as soil, water, and air. The indicator suggests a limited use of agrochemicals, promotion of organic fertilizers, and natural reproduction of seeds.

FP. (4) Low technological level: mainly associated to the human and animal work, as well as traditional techniques of sowing, harvesting, cleaning, storing, and loading of agricultural products.

4.2 Category 2: Sociopolitical dimension

The political dimension of agroecology implies the construction of agro-food alternatives through collective actions, which could be related to production, commercialization, and even political struggle [15,18] . This is achieved from a social, ecological, and social perspective, to improve sustainable development of metropolitan systems. Therefore, agroecology is considered one of the central components of the social and solidarity economy, understanding it as a model of social relations where growth or money is not placed as the central axis of the development, but the relationships between people and communities. To reach this dimension, criteria related to socioeconomic organization of agroecological practices are defined as follows:

4.2.1 *Criteria associated with sociopolitics organizations (SO)*

SO. (1) Associativity: corresponds to the community forms of organization, which are related to the decision-making processes and the management of the land.

SO. (2) Permanence: it is referred to the existence of a family or community organization that can be maintained over time through equitable relationships that contribute to social reproduction.

SO. (3) Critical posture against the agro-industrial model: it is observed in cases in which producers are threatened in their way of production, which leads them to be informed in relation to benefits of the agro-industry to subsequently defend their position.

4.3 Category 3: Urban Environment dimension

The areas in which agroecological practices are developed can be conceived from the planning perspective, as zones of enjoyment and public recreation, by allowing the mixture with other land uses such as social equipment, park areas, and others [4]. This approach promotes a more harmonic territorial development in complex urban and large scale contexts such as the metropolitan ones. The integration of agriculture in urban environments also generates important social and ecological benefits by increasing local biodiversity and improving the resilience of the urban system [5,6]. The development of agricultural activity in the city is also a source of employment and incomes, which simultaneously favors the development of local production and consumption networks. Therefore, the integration of UA within a multifunctional strategy of soil use is consistent with recent strategy of urban sustainability [13]. To guarantee these outcomes, the urban environment criteria consider:

4.3.1 *Criteria associated with MS*

MS. (1) Land use: diversity of uses, ownership of the land, and planning regulation applying over the studied territory.
MS. (2) Employment and distribution chains: corresponds to the generation of local jobs and the implementation of short production-commercialization chains.
MS. (3) Biodiversity: fulfillment of biodiverse productive goals without compromising the organization of the systems, from which the following aspects are supported: quality of soil, water, and biodiversity of the environment.
MS. (4) Resilience: in case of economic crisis or natural disasters, these areas have the capacity to provide food and water to the surrounding population.

5 ASSESING AGROECOLOGICAL PRACTICES IN CMA, CHILE

The study area corresponds to CMA, a major urban system of the Biobío Region, in central-southern Chile and second at national scale. It began to be configured in the middle of the 20th century with the industrialization of steel, which enhanced the conurbation between the regional capital, Concepción, and the port city of Talcahuano [1]. It is distinguished by its coastal and rainfed geography, flanked to the East by the Coastal Mountains. This consists of a complex water system composed of rivers Biobío and Andalién, as well as a set of lagoons and coastal wetlands. Administratively, it includes eleven (11) municipalities that house around one million inhabitants, in an area of 2,830.4 km^2, which is equivalent to 12% of the surface of the Biobío Region. In total, seven (7) of these municipalities are littorals and four (4) are located in interior areas, concentrating 81% of the total population of the region.

In order to analyze the studied territory, Regulatory Planning Schemes of the CMA were georeferenced based on QGIS 2.0 software that run in Ubuntu operating system and based on the open-code GNU/Linux platform. Lately, agricultural land with a minimal surface of 145 m^2 was mapped, based on the National Forestry Corporation cadastre from 2008. Complementing the georeference information with satellite images from Landsat, GNES/Astrium and Digital Globe, 2015 and 2016, available in Google Earth, a synthesis map was built, showing the existence of UA practices in CMA (see Fig. 1a).

Subsequently, satellite images were analyzed, experts were interviewed and field visits were carried out in order to identify invisible agroecological practices in the metropolitan area. Finally, three exemplary cases (Fig. 1b–d) were determined to be used as study elements in this research.

About the chosen study cases, the designed evaluation methodology was applied through the fulfilling of a file card for each one of the identified practices. This allowed the analysis of each dimension and criteria defined by the proposed methodology independently for each case. General results regarding the analysis of the three case studies are shown as follows:

5.3.1 *Criteria associated with the FP*

FP. (1) Obtainment of water from natural sources. All case studies are emplaced next to bodies of natural water, such as wetlands (in Case 1, Boca Sur), river (in Case 2, Cosmito), and a surface steam (in Case 3, Tome). However, only the two first cases obtain water from the natural sources, while Tome depends entirely on subsidies tank trucks and is just beginning to explore rainwater collection techniques.

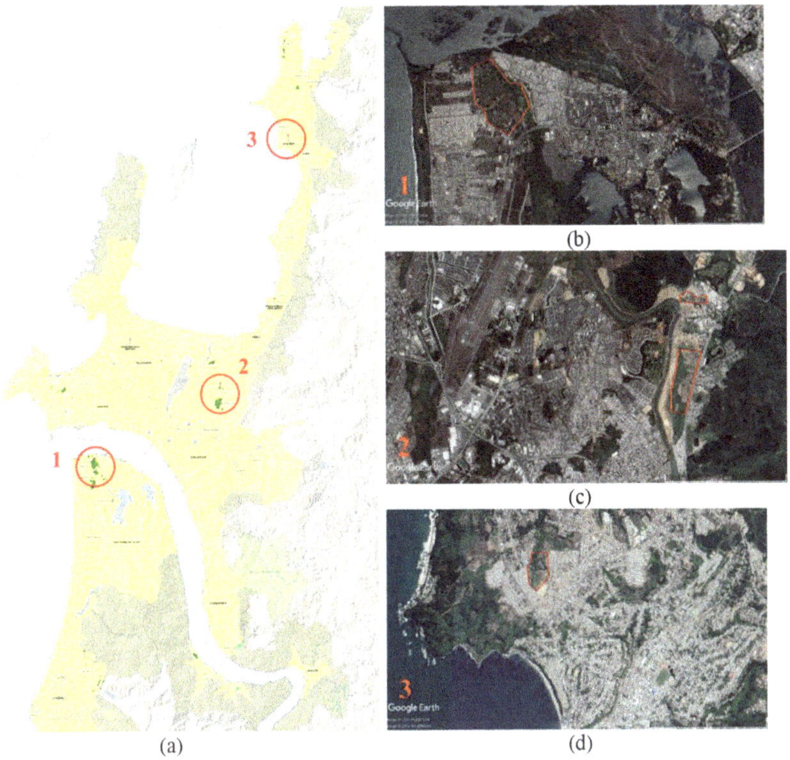

Figure 1: Map of the CMA and identification of study cases. (a) Concepción Metropolitan Area. (b) Case 1. San Pedro de la Paz, Boca Sur Sector. (c) Case 2. Penco/Concepción, Cosmito Sector. (d) Case 3. Tomé, Los Lagos de Chile Sector.

FP. (2) Small-scale production. All three case studies presented production associated to family self-consumption scale. But only Case 1 in Boca Sur and Case 2 in Cosmito perform seasonal intensive production for commercializing.
FP. (3) Low contamination and use of innocuous products. In cases 1 and 3, a scarce application of agrochemicals and low dependence on fuels was observed, either due to the high cost of inputs and because they are considered unnecessary by the producers. Case 2 in Cosmito showed a significant use of agrotoxics to improve production.
FP. (4) Low technological level. In all cases, human and/or animal work was observed, as well as simple and collective techniques of sowing and harvesting.

5.3.2 *Criteria associated with SO*

SO. (1) Associativity. Farmers participating in Boca Sur and Tome belong to formal organizations, which allow them to establish links with other organizations of local producers and associativity with community and educational centers. Case 2 in Cosmito work under the subleasing of land.
SO. (2) Permanence. The case of Boca Sur is a result of the expansion of the urban tissue over rural areas, and had been used for agricultural practices for about 180 years. The case

of Cosmito emerged as a Model Farm and has been used for agricultural purposes for approximately 120 years. In the case of Tomé, a more recent practice is observed, which is 20 years old. This was originated with the support of the NGO CET-South, with an environmental approach.

SO. (3) Critical position against the agro-industrial model. Both cases of Boca Sur and Tomé have developed diverse sociopolitical actions to protect surrounding wetland from the pressures of urban development and participate in activities of environmental training and seed bank. Cosmito case study functions under a productive logic and does not get involve in sociopolitical debates.

5.3.3 *Criteria associated to the MS*

MS. (1) Land Use. In both cases of Boca Sur and Cosmito, it is possible to identify ownership and leasing of the land. While Case 3 in Tome works on public land license by local council to be used by the social organization. Planning schemes only recognized Boca Sur as Productive Horticultural Zone (ZPH). Case 3 in Tome is regulated as Slope Lidding Zone (ZRES1). A particular situation is identified in Case 2, Cosmito, where two planning schemes overlap, regulating the land as Flood Area (ZEC3) over an Urban Extension for Residential Use (HE3).

MS. (2) Employment and distribution chains. All case studies evidence work valuation by farmers and their families, as well as fair prices in relation to the sales values in supermarkets of the surrounding. In general, the main destinations of the agricultural products were free markets around the local council area, such as Vega Monumental of Concepción, commercialized by intermediaries who buy directly from the farmers (Cases 1 and 2, Boca Sur and Cosmito). Sales to local stores are also performed by farmer's organizations, particularly by those in Boca Sur. In addition, the social organization of Case 3, Tome, sells and provide food to local institutions, such as schools and jails. Direct sales to final consumers are reduced because of being time-consuming.

MS. (3) Biodiversity: A high respect for biodiversity is observed in all practices analyzed, through the cultivation of species diversity (Case 3, Tomé), or the protection of flora and fauna of the water systems that sustain them (Case 1 and 2, Boca Sur and Cosmito).

MS. (4) Resilience: All cases studied, performed an important contribution in the event of the earthquake and tsunami of February 27, 2010. The studied organizations provided water and food to the surrounding communities. Along with this, it is known that in flood events, these areas have responded optimally, containing the floods of water courses.

6 RESULTS FROM THE APPLICATION OF THE FP-SO-MS METHODOLOGY ANALYZING AGROECOLOGICAL PRACTICES IN CMA

From field visits to each case study, together with the application of the proposed FP-SO-MS evaluation methodology, it has been possible to construct a synthesis table evidencing the contribution of the three agroecological practices analyzed in the CMA, to urban and territorial sustainability (see *Table 1*).

Out of this tool, it is possible to establish that the technical aspects associated with the FP yielded the best results within the analyzed case studies, mainly because of the small scale and low technological levels of production. SO demand a special review, yielding medium-to-high results, with Cosmito as a sensitive case study, showing a connection between associativity and critical position toward agro-industrial models. Finally, the criteria of land use associated with MS stand out with medium-to-low results, evidencing the need to improve

Table 1: Synthesis of criteria for the evaluation of agroecological practices in the CMA.

Indicator	Case 1 Boca Sur	Case 2 Cosmito	Case 3 Alto Tomé
Form of production (FP)			
FP1. Obtaining water resources	A	A	C
FP2. Small-scale production	A	A	A
FP3. Innocuous inputs	A	B	A
FP4. Low technological level	A	A	A
Sociopolitical organization (SO)			
SO1. Associativity	A	B	A
SO2. Permanence	A	A	B
SO3. Critical position	A	C	A
Metropolitan Sustainability (MS)			
MS1. Land Use	B	C	B
MS2. Employment and distribution	A	B	A
MS3. Biodiversity	A	A	A
MS4. Resilience	A	A	A

A, high; B, medium; C, low.

regulation policies and planning schemes, in order to protect the development of agroecological practices in metropolitan areas, strengthening their contribution to biodiversity, employment generation, and local resilience.

Toward the analysis of each dimension associated with the FP-SO-MS methodology, in CMA, it is possible to characterize the singularities of local agroecological practices with respect to their contribution to local sustainability.

6.1 Regarding technical-productive dimension

The strategic location inside the metropolitan area, assuring their proximity to water bodies, ensures the water self-sufficiency of the agroecological experiences analyzed. Family scale production and the use of innocuous inputs, associated to low contamination and human-scale technological levels, are elements of the agroecological practices that have shown to be contributing to the sustainability of the metropolitan system.

6.2 Regarding the sociopolitical dimension

A detailed analysis allows identifying an underlying relation between the origin of the studied initiatives, the associative structure that was adopted, and the effectiveness of the results associated to the proposed agroecology indicators.

In two of the three experiences analyzed, the base organizations have become keys for the construction of associative networks, as well as a critical attitude towards the agro-industrial model. Furthermore, because of the existence of base organizations, agroecological experiences have been recognized by local public policies and incorporated into TPIs.

6.3 Regarding the urban environment dimension

At the same time, a positive input of agroecological practices to the MS is registered. This is evidenced throughout their contribution to employment generation, biodiversity, and short production-commercialization chains. Outstanding results are registered after the earthquake and tsunami of 2010, when these productive nodes contributed to the response to the social crisis of the urban system, providing basic products to surrounding communities, after the collapse of the road network.

Regardless of this contribution to local sustainability, regulatory planning schemes have been incapable of protecting these heritage practices, which are being threatened to disappear because of road infrastructural projects and expansion of residential areas.

7 CONCLUSION: STRENGTHS, WEAKNESSES AND CRITICAL NODES OF AGROECOLOGICAL PRACTICES IN METROPOLITAN AREAS

After applying the proposed FP-SO-MS methodology on three case study areas located in CMA, Chile, the study established that the main strengths of agroecological practices in metropolitan areas are related to their location in intersticial urban spaces, together with their non-industrialized scale of production.

The emplacement in rururban periphery and interstitial spaces determines a strategic location within the metropolitan system. On the one hand, these study areas correspond to non-urbanized spaces within the Consolidated Urban Area, favoring the creation of short supply chains out of the proximity between farmers and final consumers, as noted by Soler and Rivera [13]. On the other hand, their strategic location associated to water bodies inside the metropolitan area, allow them to act as a cushion for the sustainable dumping against flooding, and mass removal, increasing urban resilience and therefore confirming Calderon [7] and Colding and Barthel's studies [14].

Simultaneously, the main characteristics of agroecological practices, such as small-scale production and low levels of contamination, in addition to a critical posture regarding preservation of the environment, have proven to be factors enhancing the contribution of agroecological practices to local biodiversity, by preserving the ecosystem in which they are inserted as Folke [5], Colding and Barthel [14] predicted.

In terms of weaknesses, land use regulation and planning schemes are critical for improving urban sustainability [13]. Therefore, appropriate zoning such as agroparks, guaranteeing the protection of local food production could be a great contribution for the sustainability and resilience of the metropolitan SES [4–6]. Additionally, developing social infrastructure and equipment would contribute to the use of these edge spaces for social encounter, as proposed by Colding and Barthel, in their Urban Green Commons proposal [14] or Segrelles in his analysis of agroparks [4].

Base organizations have shown to be critical for the success and preservation of agroecological practices, becoming a key for strengthening food sovereignty throughout innocuous low-scale of production and the construction of collaborative networks [21], thus, reinforcing the thesis of Soler and Rivera [13], and the principles of agroecological practices [16].

A closer review to the results allows identifying the potential of agroecological organizations, enhancing local work, and production throughout the reinforcing of training activities, and improving selling capabilities of the farmers.

Finally, the FO-SO-MS methodology was proven to be successful for the quick assessment of agroecological practices in metropolitan systems, allowing one to compare different case studies, as well as identifying strengths and weaknesses to be addressed, in order to enhance their contribution to local sustainability. Regarding the methodology, further analysis can be performed by specifying a new set of quantitative indicators associated with each one of the proposed criteria.

Ultimately the study corroborates the importance of bringing up to date planning schemes, as well as strengthening civil society through agroecological practices in the CMA, as agents capable to lead productive, ecological and social processes. These strategies offer the mechanism to generate adequate proposals for the sustainable organization of metropolitan areas [13].

REFERENCES

[1] Rojas, C., Muñoz, I. & García-López, M., Estructura urbana y policentrismo en el Área Metropolitana de Concepción. *EURE.*, **35(105)**, pp. 47–70, 2009. https://doi.org/10.4067/s0250-71612009000200003

[2] Casado, G.G. & Mielgo, A.A., La investigación participativa en agroecología: una herramienta para el desarrollo sustentable. *Rev Ecosistemas* [Internet], **6(1)**, 2007 Available from: http://revistaecosistemas.net/index.php/ecosistemas/article/view/135%5Cnhttp://www.redalyc.org/pdf/540/54016104.pdf

[3] Navarro, E. & Álvarez, M.E., Agroecosistemas periurbanos, un potencial latente: Contribución al análisis de la multifuncionalidad a partir de indicadores de sustentabilidad. *Revibec Rev Iberoam Econ ecológica*, **24(0)**, pp. 107–21, 2015.

[4] Segrelles, J., Agricultura periurbana, parques naturales agrarios y mercados agropecuarios locales: Una respuesta territorial y productiva a la subordinación del campo a la ciudad. *Scr,* **19(502)**, pp. 1–35, November 2015.

[5] Folke, C., Resilience: the emergence of a perspective for social-ecological systems analyses. *Glob Environ Chage* **16**, pp. 253–67, 2006.

[6] Collier, M., Nedovic-Budic, Z., Nedovic Budic, J., Aerts S., Connop D. & Foley K., Transitioning to resilience and sustainability in urban communities. *Cities*, **32(1)**, pp. S21–8, 2013.

[7] Calderón, R., El rol de las Áreas Naturales Periurbanas para la Resiliencia al Cambio Climático de las Metrópolis: El Caso de la Ciudad de México. *Rev Iberoam Econ Ecológica* [Internet], **25(0)**, pp. 69–79, 2016. Available from: http://raco.cat/index.php/Revibec/article/view/310591/400616

[8] Gomez. E.L., *Concepto De Sostenibilidad En Agroecología.*

[9] Borsdorf, A., Hidalgo, R. & Sánchez, R., A new model of urban development in Latin America: The gated communities and fenced cities in the metropolitan areas of Santiago de Chile and Valparaíso. *Cities*, **24(5)**, pp. 365–78, 2007.

[10] Galindo, J. & Giocoli, A., Los bordes de la ciudad metropolitana: apuntes para repensar la ciudad. *QRU Quad Recer en Urban* **2**, pp. 100–17, 2013.

[11] Curzio de la Concha, C., El origen y las características de los fragmentos urbano-público residuales. *Cuad Geográficos* **42(1)**, pp. 53–82, 2008.

[12] UN-HABITAT., State of Latin American and Caribbean cities. *Towards a New Urban Transition. United Nat. Naples,* pp. 1–194, 2012.

[13] Soler, M. & Rivera, M., Agricultura, urbana, sostenibilidad y soberanía alimentaria: hacia una propuesta de indicadores desde la Agroecología. *X Congr la Fed Española Sociol* pp. 1–17, 2010.

[14] Colding, J. & Barthel, S., The potential of 'Urban Green Commons' in resilience building of cities. *Ecol Econ* **86**, pp. 156–66. 2013.

[15] Sevilla, E. & Woodgate, G., Sustainable rural development: formal industrial agriculture to agroecology. In *The International Handbook of Environmental Sociology*, eds. M. Reddclift, and G.T. Woodgate, Cheltenham; 1997.

[16] Altieri. No Title.

[17] Gutiérrez, J., Aguilera, L. & González, C., Agroecología y sustentabilidad. *Converg Rev Ciencias Soci*, **15(46)**, pp. 51–87, 2008.

[18] González de Molina, M. & Caporal, F., Agroecología y política. ¿Cómo conseguir la sustentabilidad? Sobre la necesidad de una agroecología política. *Agroecologia* **8(2)**, pp. 34–43, 2013.

[19] Álvarez-Salas, L.M. & Polanco-echeverry, D.N., Ríos-osorio L., Reflexiones acerca de los aspectos epistemológicos de la agroecología. *Cuad Desarro Rural*, **11(74)**, pp. 55–74, 2014.

[20] Foro Mundial sobre Soberanía Alimentaria. Declaración de Nyéléni [Internet]. 2007. Available from: https://nyeleni.org/spip.php?article290

[21] Ortega-Cerdà, M. & Rivera-Ferre, M., Indicadores internacionales de Soberanía Alimentaria. Nuevas herramientas para una nueva agricultura. *Rev Iberoam Econ Ecológica* [Internet], **14**, pp. 53–77, 2010. Available from: http://redibec.org/IVO/rev14_04.pdf

Author index

WIT*PRESS* ...for scientists by scientists

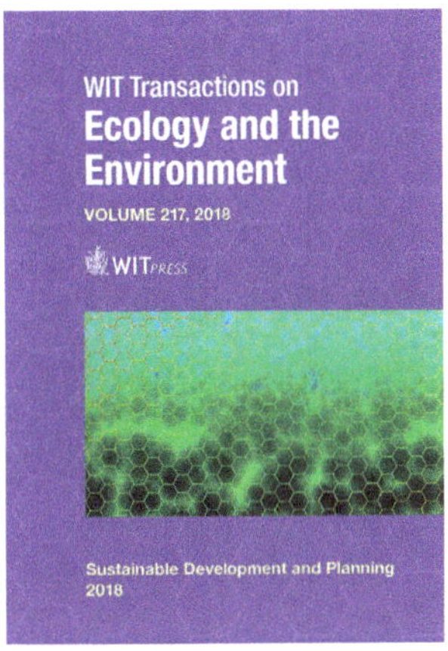

Sustainable Development and Planning X

Edited by: ***G. PASSERINI**, Marche Polytechnic University, Italy and **N. MARCHETTINI,** University of Siena , Italy*

This volume contains research from the 10th International Conference on Sustainable Development and Planning. The papers included in this volume form a collection of research from academics, policy makers, practitioners and other stakeholders from across the globe who discuss the latest advances in the field.

Problems related to development and planning, which affect rural and urban areas, are present in all regions of the world. Accelerated urbanisation has resulted in deterioration of the environment and loss of quality of life. Urban development can also aggravate problems faced by rural areas such as forests, mountain regions and coastal areas, amongst many others. Taking into consideration the interaction between different regions and developing new methodologies for monitoring, planning and implementation of novel strategies can offer solutions for mitigating environmental pollution and non-sustainable use of available resources.

Energy saving and eco-friendly building approaches have become an important part of modern development, which places special emphasis on resource optimisation. Planning has a key role to play in ensuring that these solutions as well as new materials and processes are incorporated in the most efficient manner.

The application of new academic findings to planning and development strategies, assessment tools and decision-making processes are all covered in this book.

ISBN: 978-1-78466-291-2 eISBN: 978-1-78466-292-9 Published 2018 / 1024pp

www.ingramcontent.com/pod-product-compliance
Lightning Source LLC
LaVergne TN
LVHW062240200926
848484LV00006B/1580
9781784663650